Microsoft

Table of contents

Introduction to The 2018 Update

Creating a trusted, responsible and inclusive cloud

In 2016, Microsoft launched "A Cloud for Global Good," a book of policy recommendations for governments, industry and civil society to help ensure that the opportunities of technology are evenly shared, and that challenges facing society are identified early and practical solutions applied. In the year since that launch, we have been inspired by the digital transformation journeys of countless businesses, nonprofits and governments.

In every country in which we operate, we see communities with the opportunity to flourish in new ways thanks to the power of the cloud. Cloud-powered technologies are increasingly the foundation for our lives and livelihoods, creating new businesses and business models, as well as changing the way we access services, communicate with each other and entertain ourselves. But the last year has also highlighted real challenges created by the advancement of technology. We continue to witness cyberattacks by nation-states on citizens, critical infrastructure and the institutions of democracy. We read on an almost daily basis about the criminal hacking of companies and governments to steal private and sensitive information of customers. We listen to the concerns about the loss of jobs to automation and the disruptive impact of artificial intelligence (AI) on entire sectors of the economy. And we struggle to find the right balance between freedom of expression and community safety.

There is still much work to do if we are truly going to create a cloud for global good. It is a big responsibility for every government, every business and every technology company. It certainly is a big responsibility for Microsoft.

That's why we've spent the past year engaging with leaders from governments, businesses and civil society to advance the policy framework we outlined in "A Cloud for Global Good." We have updated this book with our learnings and new ideas, building on our original policy recommendations for building a more trusted, responsible and inclusive cloud. We have also included inspiring stories of our customers who are digitally transforming and the impact they are creating for their customers, citizens and communities.

Above all, what the past year has shown us is that our optimism in the future is well placed and that the cloud will continue to play an important role in creating a better world for everyone. Once again, Microsoft is committed to partnering with stakeholders across the world to build a cloud for global good.

Brad Smith
President and Chief Legal Officer
Microsoft Corporation

Chapter 1

The role of policy in delivering a technology revolution for all

Executive summary

Humanity has come a long way in a short time. In just a few centuries, we have moved from an agrarian civilization, dispersed in small villages, to an integrated global society capable of exploring the universe and providing a high quality of life to billions of people. At the heart of this progress has been our ability to harness and distribute the productive benefits of new technology. As we move through the 21st century, this ability will prove ever more necessary if we are to address the economic, societal and health challenges that humanity faces. How do we feed a population that will reach 10 billion by 2050, and do it on less land and using less water? How do we boost productivity to accommodate shifting demographic trends that may otherwise cut economic growth in half?[1] How do we find cures for longstanding diseases and tackle a projected 70 percent rise in cancer rates over the next two decades?[2]

A new wave of technologies, built around cloud computing, offers us huge potential in our ongoing quest to build richer, cleaner, healthier societies. The foremost among these, AI, is capable of transforming the output of people and organizations across every sector. Representing an entirely new class of equipment that learns and improves over time rather than degrading, AI will augment innate human traits like creativity and sensitivity in ways that will allow us to solve previously intractable problems, boosting the productivity of our economies by up to 40 percent.[3]

We must, however, not be naïve to the challenges posed by this latest technology revolution. As with previous eras of technological change, there will be significant disruption. Businesses will be impacted as technology gives birth to new ways of working; jobs will be lost. Those who are unable to access new technology will fall behind, while those with the resources needed to build and deploy it will enjoy ever greater success. Inequality, already at record levels, could skyrocket.

In addition to these familiar challenges of technological disruption, the 12 months since launching this campaign have shined a light on the damage that results from this new technology being willfully misused. Governments have launched debilitating cyberattacks on other nation-states, disrupting critical infrastructure and interfering in the democratic process. Organized criminals continue to reap ever-increasing benefits from cybercrime, with the cost of ransomware attacks for the year up 1,500 percent from 2015 at over $5 billion.[4] As governments and companies collect ever more of our personal data, we face questions over how to ensure this data remains both secure and private.

The challenges are as real as the opportunities ... we must accompany the development of new technologies with the development and updating of new laws.

The challenges are as real as the opportunities. To address both, we must accompany the development of new technologies with the development and updating of new laws. Given the increasing pace of technological development and deployment, the urgency of our need to act grows. As with previous periods of technological change, governments must create a framework capable of delivering technology in a way that benefits and protects all. But they cannot do it alone. What follows is a suggested policy road map, designed to help governments navigate the Fourth Industrial Revolution and create an environment in which technology can be delivered in a way that is trusted, responsible and inclusive.

A Fourth Industrial Revolution that works for all

We stand on the cusp of a new era of technology, what many are referring to as a Fourth Industrial Revolution. At the heart of this revolution will be technologies enabled and underpinned by cloud computing, better known as simply "the cloud" — vast networks of distributed hyperscale datacenters, providing for the collection, storage and analysis of data at unprecedented scale and speed. It is this ability to store and process huge amounts of information that is at the heart of the data-driven technologies like AI and data analytics that will come to define our era.

It is estimated that AI alone could increase labor productivity by as much as 40 percent, driving global GDP growth by an additional 25 percent by 2035 and, in the process, generating significant material wealth for many mature and emerging societies.[5] Indeed, many now talk about AI as a completely new input or "factor of production" alongside equipment and labor. This is because although conventional equipment and buildings deteriorate over time, AI's ability to continually learn from the data it processes will enable it to become increasingly valuable. The combination of AI and other cloud-enabled technologies like data analytics is poised to drive a wider technology boom, powering advances in robotics, genomics, materials sciences and 3-D printing.

As with previous technological breakthroughs, the full potential of these new tools will only become clear over time, often in ways their creators never imagined. Look at what we have been able to achieve in combining previous inventions. Imagine, for example, explaining to Nikola Tesla, Guglielmo Marconi or Alexander Graham Bell that you'd just used inflight Wi-Fi to make a Skype call at 35,000 feet with the help of your smartphone's AI assistant.

In a similar vein, the cloud-enabled technologies that are underpinning our next technological leap will combine to help us lead richer, fuller lives in ways we can't yet fathom. According to a World Economic Forum survey of technology sector executives, 75 percent or more believe that within 10 years we'll have robotic pharmacists, automobiles manufactured using 3-D printing, and transplants of 3-D printed livers.3 If we can ensure widespread access to such technologies, it's possible to imagine a not-too-distant future in which poverty is drastically reduced, crippling diseases are eradicated, a solution for climate change is found, and new forms of communication and collaboration unleash creativity and innovation on an unprecedented scale.

But as impressive as these leaps in innovation may be, it is also possible to look at the same technological revolution and wonder if we may be headed toward a darker future. In the past 12 months we've seen companies announce plans to replace large sections of their workforce with robots.[6] The percentage of people who fear their jobs are at risk from automation is growing.[7] Moreover, concern grows that the opportunities provided by a new, digital economy are "skills-biased," or skewed toward those who have benefited from advanced education and training.

Strikingly, over the past 25 years, U.S. economic growth has produced 35 million new jobs. However, the number of jobs held by Americans with only a high school education or less has fallen by 7.3 million.[8] Similar trends are at play in advanced economies across the globe. The concentration of economic opportunities at the upper end of the income scale is likely playing a role in the increase in economic disparity, as the share of global GDP going to workers continues to decline and wealth inequalities increase.[9]

In addition to the challenge of ensuring the benefits of technological development are evenly spread, the past 12 months have also highlighted the way in which powerful new technologies can create new challenges through their willful misuse. We've seen nation-states begin to weaponize cyberspace, targeting civilian infrastructure and democratic institutions. Levels of cybercrime are rising rapidly. The number of cybersecurity breaches reported by companies increased by over 25 percent between 2016 and 2017, with the cost of ransomware attacks alone reaching $5 billion, up 1,500 percent from 2015. Governments continue to struggle to find the right balance between the need to protect public safety and preserving the right to privacy. As governments and companies collect ever more personal data, questions are raised about how to ensure this data is kept both secure and private.

The past 12 months have also highlighted the way in which powerful new technologies can create new challenges through their willful misuse.

Understandably, these challenges are raising concerns. The 2017 Edelman Trust Barometer identified a number of new inventions that are, as yet, "untrusted" by people, including driverless cars and blockchain-based currencies like Bitcoin. Nearly half of all respondents said technological innovation in general was happening too quickly and producing changes that were described as "not good."[10] Unless we find a way to ensure that everyone is able to participate in the benefits of this new technology, while also being protected from its misuse, the potential of the Fourth Industrial Revolution will be left unrealized.

Charting our way forward, by reflecting on our past

So, how should we respond? To find our path forward, it's worth reflecting on how society responded to previous periods of technological change.

The development of the cloud and the dawn of the Fourth Industrial Revolution is not the first time that a breakthrough technology has been the catalyst for profound change. By common consensus, humanity has been through three previous industrial revolutions. The First Industrial Revolution began in the U.K. in the late 18th century, when steam power led to the development of factories and machines that radically changed manufacturing and transportation.

The development of the cloud and the dawn of the Fourth Industrial Revolution is not the first time that a breakthrough technology has been the catalyst for profound change.

This was followed by a Second Industrial Revolution, beginning in the United States and built on technologies such as electricity and the internal combustion engine. Together these industrial revolutions powered the growth of modern cities, gave rise to assembly lines that changed manufacturing, and transformed transportation through the invention of airplanes, trains and automobiles. More recently, we experienced a Third Industrial Revolution that was driven by digital information processing and communications built around the core technology of the microprocessor. This ushered in the modern digital economy with the development of the personal computer, the internet and the smartphone.

Each of these revolutions delivered a huge leap forward in humanity's productive capacity. But, along the way, each was accompanied by significant disruptions for people and communities. What lessons can we learn from how society responded to previous periods of technological development?

One notable observation is that during each industrial revolution, those countries that prospered were those that created an enabling framework of laws. These laws laid the foundation for realizing the potential of new technologies while lessening their negative impacts.

The U.K., for example, which pioneered the use of coal and steam in the First Industrial Revolution, was also the first to adopt a modern system of property rights. It created the world's first patent regime in the middle of the 18th century. This was shortly before James Watt invested 12 years of his life improving Thomas Newcomen's steam engine, laying the foundation for the productivity boom of the late 18th and early 19th centuries. And with the 1833 Factory Act, which made it unlawful to force small children to work long hours in factories, and the 1863 Alkali Act, which limited the impact of coal burning on the environment, the U.K. also led the way in building out legal frameworks to begin to address the social challenges produced by the new technology. In creating these laws, the U.K. government helped build trust in these new technologies, helping ensure they were developed and delivered in a responsible way.

This dynamic was repeated during the Second Industrial Revolution, where the United States led the way in driving forward new technologies built around electricity and the internal combustion engine at the same time as making wide-sweeping interventions in education and training. 1852 saw Massachusetts create the first contemporary universal public education, quickly followed by other American states and developed countries such as the U.K., France and Japan over the course of the next 30 years. Similarly, via the 1862 Morrill Act, the U.S. was also the first to provide federal involvement in vocational colleges training people in agriculture and the "mechanical arts." These interventions both provided the updated skill base needed to power the new economy and helped ensure that a wide number of people were able to enjoy the spoils of the growing economy, by allowing them to work in those increasingly competitive industries that were deploying new technology. Reflecting on our own challenges, we must look to learn the lessons of such education programs if we are to drive more inclusive growth by delivering them over a more compressed timescale.

A history of trusted, responsible and inclusive policy

These themes of trust, responsibility and inclusion have been visible in the way that successful economies have promoted innovation during each previous period of intense technological change. Governments have time and again created laws to ensure the companies developing and deploying new technologies were accountable for their impact, driving a responsible approach to the use of new technology. The labor laws, environmental regulation and emissions standards created during the first and second industrial revolutions created better working conditions and laid the groundwork for the eventual improvements in energy efficiency and reduced emissions that companies continue to pursue today.

Similarly, educational interventions, such as the creation of universal educational systems, helped ensure that the benefits of technology are delivered in an inclusive way. In each case, successful governments are those that have been at the forefront of regulatory change.

Today, as the technology that powers our economy continues to evolve, governments must again look to update the rules, infrastructure and incentives around technology, ensuring it can be deployed in a way that benefits and protects everyone.

Indeed, given the increasing pace of technological change, there is more of an urgency to the need to review relevant legislation. It took 100 years for the steam engine to make it onto rails in the form of the locomotive, and 70 years from the invention of the telephone to its use by over 80 percent of households in the developed world. By comparison, the core technologies of our time, the personal computer and the internet, reached a similar level of penetration in less than 30 years. Indeed, the cell phone took just 15 years. The adoption of cloud services has been even swifter, with Fortune 500 companies moving from almost 0 to over 90 percent adoption in under 10 years.

We understand that governments cannot do this alone.

Given the speed at which technology is reshaping our lives and the pace at which the challenges around the deployment of new technologies continue to emerge, we must review our existing frameworks with some urgency. What follows is intended to serve as a suggested policy road map to enable governments to craft rules to help ensure that cloud-based technologies are trusted, responsible and inclusive.

We understand that governments cannot do this alone. As a company that is helping to drive technology innovation in this new era, we recognize our responsibility to work in partnership with governments and communities to help advance social and economic progress.

At Microsoft, we have made a concerted effort over the past year to play our part. To increase trust in technology, we have proposed the creation of a Digital Geneva Convention[11] to protect citizens around the world from cyberattacks. As part of our commitment to responsible technology development, we launched AI for Earth[12] to explore how our most powerful technologies can help solve some of our world's biggest challenges. And to foster inclusive access to technology, we launched the Microsoft Airband[13] Initiative to provide internet connectivity to millions of people in the United States who lack reliable and affordable broadband connections.

But we know this is just the start and that the decisions and actions we take today will affect the role that technology plays in people's lives for many years to come. We look forward to playing our role as initiator, convener, and promotor of ideas and initiatives that will help advance a cloud for global good.

Spotlight

What is the cloud?

Cloud computing — also known as simply "the cloud" — consists of networks of distributed datacenters to which users can connect their personal devices, allowing for the provision of scalable computing resources over the internet. The cloud has a number of distinguishing features when compared to conventional "on-premises" systems that people have traditionally used to do their computing.

A cloud's system of networked datacenters offers users access to **immense computing storage capacity**. Users are then able to access this storage in a scalable fashion, paying for more capacity in times when they require more storage. Think, for example, of a rapidly growing company that is able to smoothly scale the amount of computer storage it needs to hold its customer data, without having to continually upgrade its own infrastructure as it grows.

In addition to storage capacity, a cloud provider's network of datacenters — each in effect housing many individual computer servers — offers users the ability to draw on **large amounts of computing capacity**, again in a scalable fashion. This is very convenient for organizations that experience seasonality in their computing needs (think, for example, of a tax authority that has to rapidly scale up its capacity around the deadline for tax filings).

The step change in computing power provided by this ability to network servers is significant. Microsoft recently worked with the insurance firm Willis Towers Watson to calculate the cost of providing everyone on the planet with life insurance. An enormous calculation that would have taken 19 years using a conventional computer took just over 100 minutes.

It is the cloud's ability to store and process immense amounts of data with huge amounts of computing power that is at the core of data-driven technologies like AI and data analytics.

Due to the distributed and networked nature of the cloud, **users can access it anywhere**, allowing them to work, record and access data wherever they happen to be.

The shift to the cloud among enterprise organizations has been rapid. According to analyst research firm Gartner, 89 percent of companies were using cloud computing in some form by the end of 2016. Indeed, many consumers may be using the cloud today without realizing it. For example, if someone subscribes to online services such as Dropbox, iCloud, Gmail, Office 365, they are using a cloud-based service.

Cloud service models

Although numerous service models exist, there are three core types of cloud service:

Infrastructure as a service (IaaS): The provision of processing, storage, networks and other fundamental computing resources. The customer does not manage or control the underlying cloud infrastructure but has control over the operating systems, storage and deployed applications. This in effect allows customers to outsource the provision of their core computing functions, utilizing the reliability, scalability and cost-effectiveness of the cloud.

Platform as a service (PaaS): This allows the customer to create and deploy custom applications that run in the cloud using programming languages and tools supported by the provider. The customer does not manage or control the underlying cloud infrastructure but has control over the deployed applications.

Software as a service (SaaS): The provision of "off-the-shelf" applications running on cloud infrastructure. The applications are accessible from any device with an internet connection via a web browser. The most common example of a SaaS solution is cloud-based email as a service, such as the Exchange Online functionality included in Microsoft Office 365.

Chapter 2

A policy road map for a trusted, responsible and inclusive cloud

This era of rapid technology-driven change raises complex challenges and difficult contradictions. The potential for cloud computing to drive economic progress, to open the door to new ways for people to connect, to expand access to education and healthcare, and to provide new solutions to a wide range of difficult social issues is clear to almost everyone. But so are the risks and uncertainties.

The very same technologies that make this moment so promising also raise the specter of job displacement. The same tools that make it possible to launch new companies and find new ways to cure diseases can just as easily be used to commit crimes and organize terrorist attacks.

The question for government leaders and policymakers around the globe is how to harness the power of the cloud to transform people's lives for the better without unleashing the potential for dislocation and disorder. The task is a daunting one that raises fundamental questions about how to strike the right balance between competing interests such as how to protect public safety and the right to privacy; how to recognize national sovereignty without restricting the efficient flow of information across international borders; and how to provide entrepreneurs and innovators with the freedom to create and disrupt while ensuring that the benefits of change are broadly and equitably shared.

To build a cloud for global good, it will take a framework of laws that respects timeless rights and values, protects public safety, fosters innovation and the free exchange of ideas, and supports universal technology access. Governments, policy leaders and technology companies have a unique and fundamental role in helping shape this framework and deliver the unfolding technology revolution in a way that benefits all.

As highlighted in chapter one, as we move through this era of technological change — what some have called the Fourth Industrial Revolution — it is worth reflecting on how societies have responded to previous waves of technological development. What is notable about the first of the two major industrial revolutions is the way in which those countries that were at the forefront of their era's economic development were also at the forefront of exploring how to create a regulatory framework to help guide the delivery of new technology in a trusted, responsible and inclusive way.

By creating this enabling framework, governments helped build the trust in new technology necessary for it to be rolled out effectively. They were also able to help impose greater accountability on those who would create and deploy this new technology, in addition to providing the right skills and incentives to help ensure that as many as people as possible were able to benefit from this technology. If we are to make the most of the huge potential that the Fourth Industrial Revolution has to offer, we must look to emulate this approach.

... a framework for implementing a new generation of laws, designed specifically to capture the benefits of a new wave of technologies based on cloud computing.

What follows is a set of policy considerations and recommendations that offers a framework for implementing a ***new*** generation of laws, designed specifically to capture the benefits of a new wave of technologies based on cloud computing. As well as helping countries realize the upsides of this technology, they are also designed to help manage the challenges. Developed in consultation with legal experts, policymakers, industry organizations, community leaders, business users and individuals, these policy considerations are organized to reflect the underlying principles that will be essential to creating a cloud for global good — trust, responsibility and inclusion.

In the trusted cloud section, you will find policy recommendations that focus on privacy, national sovereignty and public safety. The policy recommendations in the responsible cloud section center on environmental sustainability, human rights, protecting people from the dangers of online exploitation and fraud, and AI. Policy recommendations in the inclusive cloud section include education and skills training, accessibility, affordability, and support for small businesses.

And although we believe that a new generation of laws and policies is essential given the revolutionary impact of the transformation that is underway, we also recognize that policy change takes time and often lags behind the pace of technology. We believe the technology companies that are driving these changes have an important role to play in helping policymakers anticipate the challenges that lie ahead.

So much of what makes technology work in the modern world is the creation of international standards, industry codes and government certifications. So, we remain committed to working across our industry, with our partners and governments, to find the right balance between laws, regulation and standards.

Policy section

A trusted cloud

print("mirror_ob", mirror_ob)
# put mirror modifier on modifier_ob
mirror_mod = modifier_ob.modifiers
# set mirror object to mirror_ob
mirror_mod.mirror_object = mirror
if _operation == "MIRROR_X":
mirror_mod.use_x = True
mirror_mod.use_y = False
mirror_mod.use_z = False
elif _operation == "MIRROR_Y":
mirror_mod.use_x = False
mirror_mod.use_y = True
mirror_mod.use_z = False
elif _operation == "MIRROR_Z":
mirror_mod.use_x = False
mirror_mod.use_y = False
mirror_mod.use_z = True
at the end -add back

Policy recommendation

Protecting personal privacy

The opportunity

If the cloud is the technology that will underpin the Fourth Industrial Revolution, the fuel that will power it is data. From AI, to machine learning, to data analytics, the tools that will generate the insights on which we can build healthier, cleaner, more prosperous societies are not just data-driven but data-hungry. The more data that these systems can process, the more valuable their outputs, allowing businesses to save costs and build better products and services, and allowing researchers to gain better insights into difficult problems.

The challenge

When businesses and governments hold data that people generate in the ordinary course of daily life using mobile and smartphones and other devices, it understandably creates concerns about the loss of personal privacy, raises fears about the loss of control over decisions made based on algorithms, and increases the risk that observations and predictions based on data analytics will create negative outcomes for individuals. People will be reluctant to adopt cloud services if they do not have confidence that their data will be private and secure.

Governments can establish broadly applicable binding legal norms to provide people with legal assurances, giving them confidence that their data is safe in the cloud and that businesses and governments are accountable for the fair use of advanced analytics and algorithmic decision-making.

Policy recommendations

Governments should establish clear, enforceable privacy frameworks that include strong privacy protections while enabling citizens to take advantage of the benefits of cloud computing that are dependent upon data. Privacy frameworks should provide meaningful autonomy for individuals and require organizational accountability for strong privacy protections and fair data use.

Privacy frameworks for the cloud should build on longstanding privacy principles. Chief among these is that people should have reasonable choice over whether personal data is collected and how it is used.

To enable informed decision-making, organizations must provide clear explanations about how they collect, store, use and share personal data.

These and other key principles should be reflected in laws so it is clear to technology companies how they can achieve compliance, but without government mandates for the approaches that companies should take to achieve compliance, as these may become outdated, inhibit innovation or be counterproductive.

Governments may want to consider the following goals in crafting privacy frameworks for the cloud era:

Promote transparency and control. People should have meaningful control over the use and disclosure of their personal data. To achieve this, privacy information should be provided at key points in the user experience, and people should have access to tools that make it easy to control how their data is collected and used. Where complex data analytics make simple transparency and granular user control impossible, consumers should expect higher levels of accountability from industry to help ensure the fair use of data, including plain language explanations of analytic processes and steps for remediation of unfair outcomes.

Maintain strong requirements for consent. Consent is an important legal ground for processing data, and the requirements for obtaining consent should be strong. For instance, affirmative consent should be required in circumstances where people may not reasonably expect that data is being collected or when the data being collected is sensitive and presents risks of significant privacy harms.

Permit data processing on grounds other than consent. Providing notice and obtaining affirmative consent is at times either impractical or unnecessary. Governments should consider data processing to be legitimate even in the absence of consent when processing is reasonably expected by an individual, there is minimal impact on an individual's interests and rights, or any impact on an individual's rights has been sufficiently mitigated through safeguards. Allowing for processing on such grounds is vital for enabling companies to collect data that is necessary to support, deliver and improve a variety of services for the benefit of organizations, individuals and society.

Require organizations to establish sound privacy practices. Privacy laws should require organizations to demonstrate that they have established sound privacy policies that, at a minimum, ensure compliance with legal requirements. This principle should apply to organizations that determine the purposes and means of processing data and those that process data only on behalf of other organizations. It should also apply regardless of where an organization transfers data or whether it engages other organizations to process the data.

Enable data analytics. Privacy frameworks should not be so restrictive that they prevent governments, businesses and other organizations from using data analytics to draw insights in an ethical manner. Privacy frameworks can achieve this balance by encouraging the deidentification of data sets, allowing researchers to continue to innovate but not at the expense of the personal data of specific individuals. To encourage the maximum possible privacy protection, governments should encourage the use of deidentification techniques that reduce aggregate privacy risk for individuals, even if those techniques may not be able to guarantee full, permanently irreversible deidentification of an individual.

Facilitate cross-border data flows that are protected through appropriate safeguards. As well-intentioned as laws restricting the cross-border transfer of data or data residency requirements might be, they can also be difficult to implement, have chilling effects on the economy, and be unable to address the primary privacy concerns associated with data processing. A more effective approach is to adopt regulation aligned with global standards or contracts that protect personal data regardless of its location. Such an approach can also help to improve resiliency and security and make data processing services more efficient by reducing latency. Furthermore, it should be incumbent on data processing companies to understand the laws in each country of origin and make sure that data is managed accordingly.

Evidence and further reading:

Microsoft EU Policy: EU-U.S. Privacy Shield: Progress for privacy rights

World Economic Forum report: Rethinking Personal Data: Trust and Context in User-Centered Data Ecosystems

IPAA blog: Ten Steps to a Quality Privacy Program, Part Three: Privacy By Design Tools

Microsoft On the Issues: For everyone to benefit from technology, we need to ensure the free flow of information

For links to these and other resources, please visit: http://www.microsoft.com/cloudforgood

Policy recommendation

Government access to data

The opportunity

The cloud offers us enormous potential, not only in driving innovation and efficiencies but also in serving as a secure location for storing confidential and sensitive information. These new technologies, however, raise challenging questions about the rules for how related data should be treated. Businesses and individuals have the reasonable expectation that the information they create and store in digital form should be accorded the same privacy protections as information they commit to paper.

The challenge

To fight crime and protect public safety, governments have a clear and compelling need to access digital evidence. At the same time, citizens have an expectation of due process and the rule of law, which are essential to maintaining trust in technology. This makes it a critical priority to craft modern laws that provide law enforcement and national security agencies with appropriate mechanisms to access digital evidence pursuant to lawful process. These laws should protect citizens' fundamental privacy rights and respect the sovereignty of other nations.

In addition, the rapid adoption of cloud services coupled with the corresponding rise in transnational criminal activity raises new challenges for law enforcement. But because most countries' laws have not kept pace with technology, today when information is moved to the cloud, there is uncertainty about the legal frameworks that govern access to private information.

Due to the lack of international frameworks for accessing digital evidence, governments are increasingly taking unilateral steps to seize information stored outside their border. This can create unresolvable jurisdictional conflicts that may undermine laws or force companies to choose to disregard the laws of one country in order to comply with the laws of another. Rather than circumvent established mechanisms for cross-border cooperation, governments should modernize outdated systems and, where necessary, create complementary mechanisms that operate with the efficiency required to meet today's challenges and safeguard time-honored values, including privacy and human rights.

Policy recommendations

To enable law enforcement agencies to protect public safety, governments sometimes require access to digital information, including data stored in the cloud. However, in doing so, they can undermine public trust in cloud computing.

Therefore, it is important that governments find ways to prioritize both public safety on one hand and personal privacy and freedoms on the other by adopting clear legal rules for seizing digital evidence. In developing such rules, governments should consider the following:

Allow access to digital information only pursuant to lawful process. Any framework regulating a government's ability to access digital information stored with technology providers must begin by recognizing the general principle that all access should be pursuant to the rule of law.

Ensure the right of technology providers to challenge. Technology providers should have the opportunity to challenge such process on behalf of their customers to ensure that governments are acting within the law and are respecting the rights of their users. This is a critical check on the use of government investigative powers — one that has proved effective in the United States.[13]

Require rigorous forms of legal process for more sensitive information. Technology companies store at least three types of information for their customers: (1) content, which includes information in emails and other electronic files; (2) noncontent information, which includes information relating to a user but excludes user content; and (3) subscriber information, such as identifying details about the subscriber of a service. Content is the most sensitive category of data because it contains the substance or meaning of a person's communications or documents. It is therefore appropriate to require more rigorous forms of legal process — subject to additional layers of judicial oversight — when the government seeks to access content. Though democratic governments around the world will determine their own appropriate standards, the United States' requirements of a warrant for the seizure of content — issued by a neutral magistrate based on a finding of probable cause — and a court order for transactional logs offer a model that is worthy of emulation.

Authorize disclosure in emergencies. Although governments should only be permitted to access digital information stored in the cloud through lawful process, narrow exceptions may be appropriate for emergency situations, such as when there is a reasonable, good faith basis to believe that access is needed to avoid death or serious physical injury. Such an exception can be especially crucial when law enforcement agencies face an ongoing emergency. Though the occasions when this type of exception are required are likely to remain relatively rare (as evidenced by Microsoft's annual transparency report, which includes the number of emergency requests it receives by country), such exceptions can save lives.

Support transparency. In recent years, the technology industry has secured the right to publish aggregate data about the number and types of requests it receives for digital evidence. In addition to laws that permit this level of transparency, governments should allow companies to publish detailed information (including the number of requests received and the number of customers impacted). This will help ensure that the public can understand how governments exercise their investigative authority over information stored in the cloud. Technology companies have been publishing information about law enforcement requests for years, and in 2014, new levels of transparency on the part of the United States helped demonstrate that similar information about national security requests can be made available to the public.[14]

Provide user notice. Except in limited cases, individuals and organizations have a right to know when governments access their digital information. Secrecy should be the exception, not the rule. When secrecy is required, investigators should make their case to an independent authority, such as a judge. Governments should be required to provide case-specific facts to justify any limitations on the cloud provider's ability to notify its customers of the request. And, just as important, any nondisclosure obligations imposed on a cloud provider should be limited in duration and scope to the narrowly defined objectives of the specific investigation. When necessary, cloud providers should be permitted to challenge these orders to ensure that governments operate within the law. Though it remains inadequate and in need of significant improvement, U.S. law that governs the issuance of gag orders in criminal cases is better than analogous laws in many other countries.

Modernize rules governing appropriate targets of requests for cloud data. With more and more public and private organizations moving their digital information to the cloud and many newer companies using cloud-based infrastructure to deliver applications and services to customers, governments often have multiple sources for the digital information they seek. Whenever possible, digital evidence should be obtained from the company most directly offering the service to customers. In many cases this will not be the cloud provider. Going directly to the company that is the data controller (usually the customer or consumer) can often be done without jeopardizing an investigation.

Respect international borders and sovereignty. The lack of modernized laws and international frameworks for accessing digital evidence and the increase in unilateral actions by law enforcement agencies to seize information stored outside their border threaten to erode consumer trust and are creating difficult legal situations for companies that provide cloud services. The existing mutual legal assistance process should be modernized and streamlined to ensure that it can continue to serve its purpose in a modern world. To do that, governments should develop a system that empowers law enforcement agencies to combat the many threats we face today, from terrorism to cybersecurity, while strengthening global protections for human rights and privacy and promoting the free flow of information. Important work is already being done in this area by academics and a small number of governments seeking to develop a model that can be widely replicated.

Promote trust through security. In recent years, law enforcement agencies have argued that encryption impedes legitimate investigations by putting encrypted information beyond their reach. However, some of the proposed solutions to this issue — from weakening encryption algorithms to mandating that governments be provided with encryption keys — raise significant concerns. Encryption plays an important role in protecting private data from hackers and other malicious actors. Regulatory or legal reforms in this area must not undermine security, an essential element of users' trust in technology.

Evidence and further reading:

Microsoft On the Issues: Keeping secrecy the exception, not the rule: An issue for both consumers and businesses

Reform Government Surveillance blog: RGS Statement on US-UK Data Protection Discussions

Lawfare: Cross-Border Data Requests: A Proposed Framework

Just Security: Privacy Rights Advocates Embrace DOJ's Cross-Border Data Proposal

The Guardian: Tech giants reach White House deal on NSA surveillance of customer data

Microsoft On the issues: DOJ acts to curb the overuse of secrecy orders. Now it's Congress' turn.

For links to these and other resources, please visit: http://www.microsoft.com/cloudforgood

Policy recommendation

Promoting the free-flow of data

The opportunity

In our increasingly interconnected world, the ability to transfer digital information across borders is essential to economic growth and opportunity. McKinsey Global Institute estimates that the international flow of data contributed 2.8 trillion U.S. dollars to the global economy in 2014,[15] a figure that could reach 11 trillion U.S. dollars by 2025.[16]

According to Michael Porter and James Heppelmann, writing in the Harvard Business Review, data-fueled technologies have the potential to drive a sharp increase in innovation, productivity gains and economic growth.[17] Policymakers are beginning to recognize that cloud computing is creating opportunities for companies large and small to drive innovation and transform every aspect of business operations. Access to these technologies and the freedom to use them to send data across borders is especially important for small and midsize companies because it can enable them to compete against larger businesses and reach customers around the globe in ways that have never been possible before.

The challenge

Most governments recognize that innovations powered by cloud computing offer huge potential benefits, and they understand that these innovations often require the movement of data across international borders. At the same time, there are growing concerns and misconceptions about the potential to misuse digital technologies to exploit children, commit fraud and other crimes, and carry out acts of terrorism.

Striking a balance that facilitates the smooth flow of data and provides appropriate capabilities to preserve privacy, protect individual and public safety, and promote national security is a difficult challenge. Compounding the difficulty is the fact that many existing laws and agreements governing the flow of data across international borders were created years — even decades — before the widespread adoption of email, social networks, texting and other capabilities that we take for granted today.

As a result, companies large and small face legal restrictions that sometimes limit their ability to store, transfer and process data across borders. These restrictions include legal mandates to store data locally, local supply requirements, and the effects of conflicts between laws in different jurisdictions. The impacts include higher costs, reduced economic opportunities, closed markets, and restricted access for consumers to new products and services.

Policy recommendations

Governments can help businesses and consumers realize the benefits of cloud computing without sacrificing their ability to protect privacy and public safety. Although the responsibility to create trust primarily lies with technology companies, governments have a fundamentally important role to play in encouraging greater use of cloud services to help businesses grow and deliver innovative services to consumers. As governments assert national sovereignty over online content and conduct, they must also respect the legitimate interests and sovereignty of other jurisdictions and recognize the critical importance of access to an increasingly global network of cloud services for businesses large and small.

Steps governments can take to protect access to cloud-based services that rely on cross-border data transfers and to preserve their own regulatory authority include:

Minimize disruptions to data flows in domestic legislation. Virtually all companies today use services that involve the transfer of data, and many of these transfers cross borders. When drafting domestic rules, governments should minimize adverse impacts on products or services that involve cross-border data transfers. In particular, they should avoid rules that prohibit data from being stored or processed in other jurisdictions or that require the use of domestic cloud services providers or datacenters. In some cases, such provisions are incompatible with existing international obligations.

Encourage e-commerce. Electronic commerce, which invariably involves cross-border data flows, has the potential to expand opportunity and foster equal access to the benefits of cloud computing because it brings the global marketplace to every consumer with an internet connection, while enabling even the smallest local business to reach consumers and suppliers anywhere in the world. To ensure that e-commerce reaches its full potential, governments should refrain from imposing customs duties or other taxes on cross-border electronic transmissions (consistent with the 1998 WTO moratorium on e-commerce duties) and commit to extending nondiscriminatory treatment to digital products and services.

Avoid establishing conflicting rules that raise barriers. In a world where data flows are global, the risk of conflicting national rules is substantial. Because compliance costs from conflicting rules are enormous — and may exceed what many smaller firms can afford — governments should ensure that legislation provides maximum flexibility and creates the least risk of conflict.

Adopt trade commitments that foster data-driven innovation. When negotiating new trade agreements or updating existing ones, governments should commit to allowing the free flow of data across borders and prohibiting local computing requirements, subject to appropriate exceptions. Such rules are vital in the era of the cloud, when the ability to move data freely underpins many of the most innovative products and services. They should also take steps to open up trade and investment in services, and to apply the same standards of fair and nondiscriminatory treatment to digitally delivered products and services as to those in the physical world. With respect to intellectual property, governments should provide meaningful protections for rights holders as well as safeguards to foster the internet's continued growth as a platform for free expression, innovation and digital commerce. Governments should provide online service providers with a "safe harbor" from liability for infringing or otherwise unlawful content posted by third parties, and should embrace balance in copyright law by adopting flexible exceptions and limitations that encourage innovative activities such as data analytics, machine learning, and commercial text and data mining of lawfully accessible content.

Evidence and further reading:

World Bank: World Development Report 2016: Digital Dividends

Information Technology & Innovation Foundation: Cross-Border Data Flows Enable Growth in All Industries

European Centre for International Political Economy (ECIPE): The Costs of Data Localization: A Friendly Fire on Economic Recovery

McKinsey Global Institute: Digital Globalization: The New Era of Global Flows

Organisation for Economic Co-operation and Development (OECD): Economic and Social Benefits of Internet Openness: 2016 Ministerial Meeting on the Internet Economy—Background Report

BSA | The Software Alliance: What's the Big Deal with Data?

For links to these and other resources, please visit:
http://www.microsoft.com/cloudforgood

Policy recommendation

Ensuring secure and reliable infrastructure

The opportunity

As cloud computing gives rise to powerful new capabilities, it offers the potential to increase productivity and innovation, reduce costs, and drive new levels of security and resiliency. The last two are particularly important, as the ever-increasing connectivity — of both devices and people — has created new ways for malicious actors to attack or commit crime against people. To be effective, our online defenses need to embrace the security advancements offered by cloud computing — from scalability and geographic replication to the use of machine learning and other innovations.

The challenge

Cloud computing represents a seismic shift from traditional computing — not just in what it enables, but in how it is built, managed and used. To address the risks and threats of the cloud computing era, governments will need to adapt existing security programs and policies and enhance current approaches to ensuring the security and resilience of their systems.

The shift will require not only much closer cooperation with cloud vendors to ensure the security outcomes governments are seeking are met effectively, but also a change in how the regulatory landscape is managed. Sectoral and vertical approaches to critical infrastructures will have to be reassessed, as the technology underpinning them cuts across them horizontally. Moreover, the global nature of security threats will make cross-border partnerships and harmonized legal approaches even more important.

Policy recommendations

Governments must play a central role in developing, evolving and implementing security policies. Effective approaches will not only increase domestic and global security, but also enable continued innovation, productivity and economic opportunity. Less effective approaches will create heavy operational costs without realizing the intended and much-needed security benefits.

To create efficient policy and regulatory frameworks that ensure secure and reliable services are used, the following steps are recommended:

Establish risk management processes and prioritize efforts. Any regulations that are introduced should be based on a thorough understanding of the threats, vulnerabilities and potential consequences facing the country. In adopting a risk-based approach, governments will recognize that all activities involve some degree of risk and that no organization has unlimited resources to apply to security. The approach will also allow governments to prioritize their security investments on the most important national assets, ensuring that these have sufficient protections in place.

Implement a data classification system for the cloud. Data classification is the process of dividing data into distinct categories based on sensitivity levels and risk profiles, and then articulating the security controls needed for each level to manage risks appropriately. Having a cloud-specific data classification system will help enterprises and government agencies identify both their most sensitive and least sensitive materials and evaluate the costs and benefits of storing varying levels of sensitive materials in the cloud. To the extent possible, governments may adapt existing data classification schemes to data stored in the cloud as evidenced by the approach of the U.K. government's classification program.

Establish public-private partnerships. Public-private partnerships are a cornerstone of effectively managing security risks in both the short and long term. They are essential for boosting trust among and between the operators and the government. Their focus areas could include coming to an agreement on common cybersecurity baselines; establishing effective coordinating structures and information-sharing processes and protocols; identifying and exchanging ideas, approaches and best practices for improving security; and improving international coordination.

Set baseline security measures for government and critical infrastructures. Security baselines are a foundational set of policies, outcomes, activities, practices and controls that help manage cybersecurity risk. They can take the form of voluntary guidance, coupled with incentives (e.g., procurement requirements or tax subsidies), or be implemented through a mandatory regulatory requirement, in particular where an elevated need for assurance arises from the risk environment. Governments should consider utilizing existing best practices, such as the National Institute of Standards and Technology's (NIST) Cybersecurity Framework, to ensure speedy adoption and international harmonization.

Develop outcomes-focused frameworks. It is essential that any regulations introduced are outcomes-focused, articulating what organizations should aim to achieve (e.g., "control logical access to critical resources") rather than how organizations should implement security (e.g., "utilize two-factor authentication"). In the rapidly changing world of cybersecurity, prescriptive approaches will quickly become out of date or leave the country out of step with international best practices. Moreover, outcomes-focused approaches allow for variability in the architecture of technology and give the user the flexibility of using whatever best fits their needs.

Develop a common security compliance model for critical information infrastructures. Because every sector of the economy depends on technology, there is a high degree of commonality of the risks and associated controls and policies across the different sectors. Rather than developing minimum security goals and standards for each individual sector, government should seek to harmonize approaches by developing an overarching security compliance model for critical information infrastructures. To ensure specific risks are addressed, governments should also allow individual sectors to establish a smaller subset of additional requirements appropriate for their unique operating environments.

Leverage global standards in national cybersecurity efforts. The threats to cyberspace do not stop at national borders. It is therefore essential that governments adopt approaches for encouraging cybersecurity that acknowledge that reality. National approaches should therefore integrate international standards to the maximum extent possible, keeping the goal of harmonization in mind. Moreover, by leveraging global standards as the basis of their certifications, governments can improve efficiency, lower costs and improve market competition.

Evidence and further reading:

Microsoft white paper: Transforming Government: Cloud Policy Framework for Innovation, Security and Resilience

Microsoft white paper: Transforming Government: A Cloud Assurance Program Guide

National Institute of Standards and Technology: National Institute of Standards and Technology (NIST) Framework for Improving Critical Infrastructure Cybersecurity

Microsoft EU Policy: Protect, respond, collaborate, deter: a new opportunity for European cybersecurity

Microsoft EU Policy: Progressing from Padlocks: Securing Industry in the Cloud

For links to these and other resources, please visit:
http://www.microsoft.com/cloudforgood

Policy recommendation

Creating a Digital Geneva Convention

The opportunity

As the transformational power of cloud computing comes into focus, there are growing concerns about the rise of cyberspace as a battlefield. Effective cybersecurity is critical to international peace and economic stability; however, governments continue to invest in greater offensive capabilities in cyberspace, and nation-state attacks on civilians are on the rise. There is a growing urgency to develop new international rules to protect civilians from nation-state threats in cyberspace, in particular in times of peace.

In short, the world needs a Digital Geneva Convention.

The challenge

The process of creating the Digital Geneva Convention involves formidable challenges and will require political will and commitment from government leaders around the world. Some important foundations have already been put in place upon which we should look to build, including, for example, the rules and principles proposed in 2015 by the United Nations Group of Governmental Experts on Developments in the Field of Information and Telecommunications in the Context of International Security.

More work remains to be done, however, to further define the cybersecurity norms that have been already agreed upon, as well as to put forward new proposals. Input from the global information and communication technology (ICT) industry is critical to ensuring that the language of diplomacy accurately reflects the realities of defending technology users at global scale.

Furthermore, work is needed to advance transparency of, and accountability for, state behavior in cyberspace. Any successful implementation will require new mechanisms for dealing with politically sensitive allegations such as attribution. Governments and the private sector need a forum where they can provide evidence to support technical attribution and obtain validation through rigorous peer review. A pragmatic and flexible path to deliver that vision should be identified.

Policy recommendations

Although there are signs of alignment around a small number of international cybersecurity norms, more progress is needed. Process continues to be slow and does not keep pace with technological advances. If we are to avoid the potentially catastrophic effects of cyberwarfare, continuous engagement among relevant actors is essential, and governments must build international rules as they have done in other contentious areas of policy that touch upon sensitive geopolitical issues.

By building on the work done to date, governments, the technology sector and civil society groups can pave the way for a legally binding agreement that will ensure a stable and secure cyberspace. The key clauses at the center of this Digital Geneva Convention should commit states to:

Refrain from attacking systems whose destruction would adversely impact the safety and security of private citizens (i.e., critical infrastructures, such as hospitals and electric companies).

Refrain from attacking systems whose destruction could damage the global economy (e.g., integrity of financial transactions) or otherwise cause major global disruption (e.g., cloud-based services).

Refrain from hacking personal accounts or private data held by journalists and private citizens involved in electoral processes.

Refrain from using information and communications technology to steal the intellectual property of private companies, including trade secrets or other confidential business information, to provide competitive advantage to other companies or commercial sectors.

Refrain from inserting or requiring "backdoors" in mass-market commercial technology products.

Agree to a clear policy for acquiring, retaining, securing, using and reporting of vulnerabilities — that reflects a strong mandate to report them to vendors — in mass market products and services.

Exercise restraint in developing cyberweapons and ensure that any that are developed are limited, precise and not reusable. States should also ensure that they maintain control of their weapons in a secure environment.

Agree to limit proliferation of cyberweapons. Governments should not distribute, or permit others to distribute, cyberweapons, and should use intelligence, law enforcement and financial sanctions tools against those who do.

Limit engagement in cyberoffensive operations to avoid creating mass damage to civilian infrastructure or facilities.

Assist private-sector efforts to detect, contain, respond and recover in the face of cyberattacks. In particular, enable the core capabilities or mechanisms required for response and recovery, including computer emergency response teams (CERTs).

Evidence and further reading:

Microsoft On the Issues: Norms are a valid and timely option for building global cybersecurity and for making progress toward a Digital Geneva Convention

Microsoft On the Issues: We need to modernize international agreements to create a safer digital world

Policy recommendation

Preventing cybercrime

The opportunity

The combination of expanded access to the internet, the explosive increase in connected devices, and the rapid expansion of innovative cloud-based services is creating tremendous economic and social opportunities for consumers, governments and businesses.

The challenge

Today, governments are struggling to confront the growing threat, sophistication and prevalence of cybercrime. Increasingly, these crimes are committed by organized groups operating in one country that target victims in another. The cross-border nature of cybercrime complicates enforcement, and inadequate legal frameworks in some countries have created safe havens for cybercriminals. The financial impact of all this cybercrime is large and growing. In 2015, the British insurance company Lloyd's estimated that cyberattacks cost businesses as much as U.S. $400 billion per year,[18] an amount that is expected to increase in the years ahead.

In addition to these economic costs, there are less tangible impacts including lost confidence in internet commerce, the erosion of individual privacy and diminished trust in online services. Each of these effects threatens to slow adoption of cloud-based innovation and reduce the benefits of promising new technologies.

Policy recommendations

Harmonization of cybercrime laws around the world combined with initiatives to facilitate faster and more effective coordination between law enforcement agencies is essential. These efforts can be pursued in an environment where each country respects the sovereignty of other nations, and where the fundamental rights and liberties of citizens are fully respected. To strengthen enforcement in a balanced way, governments should consider the following steps:

Support strong enforcement and balanced rules. To fight cybercrime effectively, law enforcement and industry must have the legal tools necessary to pursue cybercriminals wherever they are. Governments should work to update their criminal laws so that they are capable of addressing both existing and emerging threats posed by online criminals. At the same time, these laws should be consciously framed not to adversely affect innovation or the adoption of new technologies. They should also support efforts at industry self-regulation.

Adopt laws that are consistent with broadly accepted international conventions. The Council of Europe's Budapest Convention provides a good model for cybercrime legislation that can help harmonize laws and drive better cooperation across borders. Such international coordination and cooperation will help eliminate safe havens for malicious actors and minimize the risks that arise when intermediaries and other innocent parties are subject to conflicting obligations or liabilities.

Facilitate information sharing. In some cases, companies with information about online crimes face potential liability under privacy, data protection or other laws if they voluntarily share that information with law enforcement. To facilitate and encourage timely cooperation, governments should clarify rules for how companies share data with law enforcement. Lack of clarity about rules for information-sharing and liability risks may prevent companies from working with law enforcement agencies, even when cooperation could be critical to preventing or responding to cybercrime. In addition — as described in the recommendations for government access to data — enhancing the procedures and mechanisms for international, cross-border cooperation by modernizing mutual legal assistance processes will help streamline enforcement efforts and help clarify important issues related to jurisdiction and access to evidence.

Develop new ways to prevent cybercrime. Current efforts to enforce laws against cybercrime are woefully inadequate given the enormity of the problem. New approaches to going after the criminals are needed. One example may be a pilot program launched by the City of London police in partnership with private law firms using civil courts to seize cybercriminals' assets. Finding other ways to scale enforcement efforts will be critical.

Work with industry on best practices and emerging issues. Governments can take advantage of the expertise and resources of the private sector in the fight against cybercrime. Opportunities include working with industry to educate enforcement officials about new and emerging threats that technology suppliers experience in the real world and that their customers see as priorities. Governments often lack sufficient resources to deal effectively with cybercrime. Working with the private sector can help them achieve greater success, which will help drive trust in online computing.

Evidence and further reading:

Convention on Cybercrime (Budapest Convention)

The Guardian: Police to Hire Law Firms to Tackle Cyber Criminals in Radical Pilot Project

For links to this and other resources, please visit:
http://www.microsoft.com/cloudforgood

Policy section

A responsible cloud

Policy recommendation

Protecting both human rights and public safety

The opportunity

Social networks, mobile apps and other cloud services are enabling new ways to create, communicate, publish, and access news and information. These services have unlocked opportunities to strengthen freedom of expression. Societies benefit in innumerable ways, including a more informed and engaged citizenry, expanded economic opportunities, and more connected communities.

The challenge

Although technologies have evolved, time-honored values endure. International human rights laws have long recognized freedom of expression and freedom from arbitrary or unlawful interference with privacy, family, home or correspondence. Both freedoms are key contributors to human dignity and the development of human potential. Of course, any technology, whether the printing press or the cloud, can be misused to disseminate illegal or harmful content. This raises important questions for governments, communities, cloud service providers and other stakeholders, who seek to ensure personal privacy, freedom of expression, and the right to receive and impart information on the global internet while protecting public safety. As societies seek to protect human rights while combating terrorism and extremism, it is important to recognize that public safety and human rights are complementary values that reinforce each other.

Policy recommendations

Governments should adopt clear laws and regulations that are interpreted and administered under the rule of law, including advancement of international human rights laws and norms. This will enable governments to protect freedom of expression, privacy and public safety while continuing to support robust exchange of ideas and information to fuel the benefits that technology can bring to societies and economies.

In particular, governments should consider the following principles:

Adhere to the rule of law. In regulating online content, conducting surveillance or accessing data, governments should fully commit to the rule of law. This means ensuring that laws and regulations and their enforcement are transparent and respect international human rights laws and norms. Governments should be open and engage their citizenry in public debate on the enactment of laws and regulations regarding restriction of online content, governmental surveillance and access to online data. Citizens should determine how such laws and regulations will be enforced. Rule of law requires that enforcement orders and decisions be subject to independent judicial approval and review, with meaningful and trusted opportunity for companies and individuals to appeal judicial approvals or decisions. Adherence to rule of law will serve best to ensure that the benefits of cloud computing lead to human development and economic advancement.

Adopt a principled approach to online content regulation and protect freedom of expression and access to information. One of the fundamental roles and responsibilities of governments is to protect public safety. This sometimes requires the regulation of online content. Any governmental restriction on freedom of expression should respect the norms established by international law: legality, necessity and proportionality. Restrictions should be duly enacted by law, should be the least restrictive means possible and should be proportionate to the legitimate objective. Governments should ensure that laws are strictly limited to the protection of public safety, and do not prevent broad sharing of ideas — even ideas that are unpopular.

When governments demand that online service companies remove content, they should do so transparently. These demands should be made pursuant to laws and regulations that clearly define what constitutes illegal content and the types of services that must remove it. Laws and regulations should require that legal orders for the removal of illegal content be specific, narrowly tailored and sufficiently detailed to enable companies to identify precisely which content must be taken down. Such laws and legal orders should not require companies (directly, or indirectly through intermediary liability or other pressures) to proactively monitor content or make independent determinations of illegality. Laws and regulations should not restrict companies from informing the public about removal demands from governmental authorities.

Protect rights to privacy online. Governments have long accepted the responsibility to ensure that they conduct surveillance or access information only when necessary. Like freedom of expression, surveillance and access to data must be subject to rule of law, grounded in legality, necessity and proportionality. See Government access to data for more recommendations.

Respect national sovereignty through international cooperation. Given the transnational nature of the global internet, demands to remove content or disclose digital evidence will often affect other jurisdictions. Unilateral demands or actions risk violation of other countries' sovereignty, conflict of laws among nations, and potential interference with the exercise of fundamental rights. Governments should focus on strengthening international cooperation and adhering to international norms in considering content regulation, surveillance or access to data on the global internet. Where existing rules or processes for cross-border cooperation are outdated or cumbersome, governments should work together to update them so they keep up with new technologies, are adequate to address new challenges and protect human rights. Self-help is never the best option.

Noninterference with technology companies' terms of use. Online services that permit end users to post content for viewing by others usually include terms of use (aka terms of service). These terms of use are designed to advance the service provider's legitimate business purposes for the service in question, including generation of experiences appropriate to the nature of the service and the user communities they serve. Companies generally provide processes for users or others to report content that may violate the terms of use, and have procedures for review and removal of content that violates applicable terms of use. Governments should not pressure companies to change their terms of use or interfere with the way they are enforced.

Evidence and further reading:

Microsoft On the Issues: Microsoft's approach to terrorist content online

Microsoft On the Issues: The importance of reporting concerns about online content and conduct to tech companies

Microsoft On the Issues: New Microsoft study: Parents have the greatest impact on young people's safety online

Microsoft On the Issues: Four in 10 US teens seek help from friends about online woes, new Microsoft study shows

Microsoft On the Issues: Facebook, Microsoft, Twitter and YouTube provide update on Global Internet Forum to Counter Terrorism

Microsoft On the Issues: Microsoft partners with Institute for Strategic Dialogue and NGOs to discourage online radicalization to violence

For links to this and other resources, please visit:
http://www.microsoft.com/cloudforgood

Policy recommendation

Reducing technology fraud and online exploitation

The opportunity

Cloud computing is revolutionizing how people work, learn, interact and play. As technology becomes an increasingly prominent part of many aspects of everyday life, more people are being brought into contact with it, whether they are children using educational technologies to learn in new ways or older people using technology to stay in contact with loved ones and access information more quickly and efficiently.

Social networks are another example of the way that new technologies are being used to connect people of all ages, allowing us to form new communities and connect with friends and colleagues. In addition, cloud services have become important vehicles through which people and governments advance fundamental values including freedom of expression, civic engagement, privacy and free access to information.

The challenge

At the same time, online services have given rise to new risks and new potential for harm, especially for vulnerable populations such as children and the elderly. Predators can use connected technologies to create and distribute images of child sexual abuse and to solicit minors for sexual exploitation. Scammers trick people into believing that they have nonexistent malware or viruses on their computer and into paying for unnecessary tech support services.

Unfortunately, the methods that criminals deploy are becoming more sophisticated and harder to detect. Cybercrime is truly borderless, making investigation and enforcement across jurisdictions very difficult.

Policy recommendations

The unique challenges of protecting children, the elderly and other vulnerable populations require a coordinated and comprehensive response. In many cases, existing laws need to be updated to address current technologies and threats, balanced against the need to protect freedom of expression, individual privacy and vibrant innovation. These updated legal frameworks should promote industry best practices and the development of technology tools that consumers can use to help protect themselves. Some of these areas include:

Strengthen and enforce laws to deter creation of online exploitation and fraud. Many existing laws that are intended to fight fraud and the exploitation of minors were not written to address online crimes and, as a result, are not vigorously enforced. According to the International Centre for Missing and Exploited Children, 35 countries still do not have legislation that deals specifically with child sexual abuse images.[19] Of the 79 countries that do, 60 do not define child sexual abuse specifically and 26 do not address computer-based offenses. In addition, many laws criminalizing the creation and distribution of images of child sexual abuse are ill-suited to the new tactics of tech-savvy child predators. In parallel with other cybercriminals, child sex offenders use defensive forensic measures including anonymization and encryption of their online illegal activities to evade law enforcement. As governments update their laws to tackle these new threats, they should work closely with child rights, advocacy and support groups, as well as technology suppliers — all of which play a role in protecting children and families in the digital age.

Support public-private partnerships. Public-private partnerships are essential to address the increasing variety and complexity of online threats. Governments, technology companies and online service providers should work together to develop and share technology tools and expertise, conduct awareness campaigns, and educate the public about online risks. To address child protection, governments should consider joining the WePROTECT Global Alliance to End Child Sexual Exploitation Online, a coalition that includes 70 countries along with technology companies and civil society organizations dedicated to eradicating child sexual exploitation and abuse online.

Promote international cooperation. Increasingly, online crimes involve perpetrators in one country and victims in another, which can hamper effective prosecution. For example, live-streaming child sexual abuse often involves victims in Southeast Asia with abusers in Europe and North America across encrypted platforms. Countries committed to addressing online child exploitation should join the Virtual Global Taskforce, which shares a common understanding that protection is often only achieved by global, not merely national, methods. Similarly, the International Mass Marketing Fraud Working Group brings together law enforcement agencies committed to cooperatively addressing mass marketing fraud (including tech support scams) that harm millions of individuals across multiple countries. These voluntary initiatives are important, but it is clear that more needs to be done. New international agreements and modernized mutual legal assistance treaties are needed to strengthen cross-border cooperation, information-sharing and enforcement.

Promote consumer education. Many online crimes can be avoided if people are better informed about how to identify threats and protect themselves. According to a Microsoft-supported survey, one in five customers have experienced a fraudulent interaction online.[20] Millennials are particularly vulnerable to fraudulent email and intrusive pop-up advertisements. Alongside NGOs and the private sector, law enforcement should focus on consumer online safety education to help people identify threats and protect themselves. Education is important at every age, and can be tailored appropriately.

Support industry self-regulation. Even as governments work to address the risks associated with online services, they can promote an environment of technological innovation and industry self-regulation that provides the flexibility to respond to the rapidly changing nature of online threats, which can be difficult to achieve solely through legislation. Governments and industry should work together to establish safety principles, and service providers should be given the opportunity — and the responsibility — to determine the means of implementation.

Evidence and further reading:

Europol's Internet Organised Crime Threat Assessment (IOCTA) 2016

WePROTECT Global Alliance to End Child Sexual Exploitation Online: http://www.weprotect.org/

STOP. THINK. CONNECT.: https://www.stopthinkconnect.org/

Virtual Global Taskforce: http://virtualglobaltaskforce.com/who-we-are/member-countries/

Microsoft On the Issues: The importance of reporting concerns about online content and conduct to tech companies

Microsoft On the Issues: New Microsoft study: Parents have the greatest impact on young people's safety online

Microsoft On the Issues: Four in 10 US teens seek help from friends about online woes, new Microsoft study shows

For links to these and other resources, please visit: http://www.microsoft.com/cloudforgood

Policy recommendation

Promoting environmental sustainability

The world's population is set to reach almost 10 billion by 2050, putting significant pressure on the world's already constrained resources. Simply feeding this population will require a huge increase in productivity, with farmers needing to produce more 70 percent more food on less land, using less water and with a more precise application of fertilizer and inputs. We also face significant environmental challenges in relation to developing sustainable fuels, addressing the impact of our activities on the climate and maintaining the biodiversity of our ecosystems.

The opportunity

The computational potential of cloud-enabled, data-driven technologies like AI and data analytics provide fantastic potential in helping us address these challenges. By using the cloud to generate, gather and analyze data about our world, we will be able to discern new trends and design new solutions to help us build a cleaner, healthier world in which to live. Microsoft is excited about the potential of these tools and has funded the AI for Earth initiative with a view to helping explore the full range of possibilities re tech and our environment.

The challenge

But these new technologies also create their own challenges. As we continue to build out the datacenter infrastructure that lies at the heart of our global cloud, we use more energy. As one indicator of this energy use, Microsoft's cloud activities consume nearly as much electrical power as a small American state, and in the next five years are estimated to use as much as a small European country. But these datacenters consumer more than energy; they also take up land and water. To operate in an ever more sustainable way, cloud providers continue to invest in research, development and infrastructure to harness the benefits of new clean technology, deploying new energy-efficient datacenter and building technology, using water risk monitors, and sourcing our energy from renewable sources.

The challenges that we face as a global company to operate in an ever-more sustainable way are the same challenges being tackled by governments across the world. One important aspect is the fundamental complexity of expanding the availability of clean energy. Each country has its own energy policy framework, fuel mix, market structure, legacy infrastructure and political priorities. In some countries, energy infrastructure is outdated and integration with renewable sources is difficult. In others, better alignment of policies with consumer demand for clean energy can facilitate direct purchasing and on-site generation of renewable energy and encourage greater transparency and competitive pricing that will help datacenters and their customers use energy more efficiently.

Beyond energy, countries must work together and take a modern approach to tackling the major challenges around the impact of our activity on our environment. This includes addressing issues around the warming of our climate and the impact of a growing population on our land and marine environments. In meeting these great challenges, we should look to use the best technologies at our disposal. As leading companies around the world utilize data-driven cloud technologies to improve their productivity and efficiency, we must also use these tools to help us improve our environment. This will require us to generate, collect and analyze more data about how our world works and how we interact with it, as well as opening up the significant amounts of data we already collect for deeper analysis.

Policy recommendations

Policies that promote sustainable practices and support renewable and clean energy are necessary to help address our environmental challenges. In addition, such policies can also provide countries with a competitive advantage in attracting technology firms and other investments that prioritize clean energy sourcing. As with other areas of the technology policy landscape, broad partnerships and collaboration between governments, businesses and nongovernmental communities will be essential to progress.

Steps that can expand access to clean energy and increase energy and resource efficiency include:

Increasing access to clean energy. Energy sources, infrastructure and policies vary greatly from country to country. In nations where renewable and other clean energy are viable options, governments should facilitate the development of new renewable energy sources by setting targets and providing incentives that promote the development and use of clean energy. Although policy design will vary by country, policy options may include renewable portfolio standards that require a certain amount of electricity to be generated by solar, wind, hydro and other zero-carbon sources; tax incentives for renewable and other cleaner energy; and pollution rules that encourage a shift to cleaner energy sources. In countries where these incentives already exist, governments can accelerate clean energy development by allowing direct energy investment by large consumers either on-site or through third parties and by facilitating partnerships between consumers and utilities to increase the availability of renewable energy in a cost-effective way.

Encourage energy efficiency and research and development. Because cloud-based services are generally more efficient than individual servers and datacenters, increased use of the cloud can drive energy efficiency gains — a recent study found that organizations can reduce energy use by 30 percent to 90 percent when they move from on-premises software to the cloud.[21] Governments can encourage these gains through policies and regulations that encourage migration to the cloud. Ongoing research and development is also critical to improve energy efficiency and develop new clean energy technologies. Governments should encourage investment in research and development and support public-private partnerships, particularly in new battery technologies that can store clean energy at scale and smart-grid technology that can use real-time information to balance power distribution. In addition, cloud services provide valuable tools for improving efficiency, such as reduced energy use in water management, public transportation and residential heating. Governments, academic research institutions and companies should continue to work together to use cloud computing technologies to discover and drive these kinds of efficiencies.

Promote transparency. Governments should increase transparency on pricing and consumption of energy and resource use. Government rules and incentives that increase public access to information about when and where energy demand is high, when and where energy supply is plentiful, and what sources are the most efficient, clean and affordable can reduce energy waste and accelerate clean energy development. The cloud can help the energy and other resource sectors organize, analyze and deliver this type of information so that producers and consumers can make more informed decisions and create more efficient operations. In addition, specific policies that require public reporting and transparency of energy consumption by buildings can help reduce energy use.

Accelerate energy efficiency through smart energy systems. The use of cloud-based smart systems for buildings and urban infrastructure can significantly reduce energy consumption. To effectively use these smart systems, governments should promote wide-scale broadband connectivity and the deployment of smart devices either directly or through supporting policies.

Open up existing data sets for analysis. Data-driven technologies like AI represent the most advanced analytical and problem-solving tools that humanity has to draw on. They represent immense potential in understanding and responding to our environmental challenges. They are, however, powered by data. The more data that is fed in, the more valuable the outputs. Governments should look to make environmental and other relevant data sets more readily available for analysis. The steps taken by organizations such as the U.K.'s Department for Environment, Food and Rural Affairs to make 8,000 data sets freely available are a positive lead to follow.

Fund the creation of new data sets for analysis. In addition to opening up currently available data, governments should continue to upgrade the capacity of our earth observation systems for sky, land and water environments. In addition to generating the additional data necessary to most effectively tackle our environmental challenges, the insights gained may have the potential to deliver additional scientific and commercial value.

Evidence and further reading:

Brad Smith, LinkedIn post: Microsoft Reaction to the White House Announcement on the Paris Agreement

Microsoft News Center: Microsoft, Accenture, and WSP Environment and Energy Study Shows Significant Energy and Carbon Emissions Reduction Potential From Cloud Computing

Microsoft Transform: Ecolab and Microsoft team to face water shortage challenges

Microsoft Green: How Microsoft Technology is Enabling an Autonomous Grid

Microsoft Green: Protecting Biodiversity with Artificial Intelligence

Microsoft Green: Microsoft's Redmond Campus: 500 Acres and "Zero Waste"

Microsoft Green: Energy Efficiency and Designing the Datacenters of the Future

Microsoft On the Issues: Greener datacenters for a brighter future: Microsoft's commitment to renewable energy

For links to these and other resources, please visit:
http://www.microsoft.com/cloudforgood

Policy recommendation

Amplifying human ingenuity through artificial intelligence

The opportunity

The ways in which AI can augment human capabilities to improve society are almost limitless. Indeed, with the advent of self-driving cars, digital personal assistants that can anticipate our needs, and computerized health diagnosis, AI is beginning to change people's lives for the better. The promise of AI is that knowledge gained from applying analytics to the wealth of data that is available today will enhance any decision-making process with additional intelligence, helping us produce quicker, more effective outcomes.

Recent research by Accenture estimated that AI could double annual economic growth rates by 2035, and boost labor productivity by up to 40 percent. A study by McKinsey suggests that AI-enabled digital platforms that match workers to task-based jobs could add 2 percent to global GDP and create the equivalent of 72 million full-time jobs by 2025. And the United Nations has embraced AI as critical to the achievement of the UN's Sustainable Development Goals.

This technology holds much promise for the future as well as providing much opportunity for us today.

The challenge

Although the potential for AI to improve lives is enormous, not everyone is confident that the emergence of machines that can perceive, learn and make recommendations necessarily promises a better future. Big questions exist, and there are much more immediate questions about what AI could mean for jobs and income inequality. A World Bank report suggests that nearly 60 percent of jobs in developed countries could be automated in the coming years. And the rapid emergence of AI raises policy and ethical concerns that touch on issues ranging from privacy and transparency to fairness and safety.

These concerns are understandable. The questions that arise are more than just reasonable; finding the right answers will be essential if we are going to create the levels of trust that will be required for people to adopt AI on a global scale and ensure that the benefits of AI are broadly shared and the ultimate potential for AI to improve human lives is achieved.

Policy recommendations

To foster innovation in AI and the implementation of AI capabilities, governments should create legal and policy frameworks that enable access to data, encourage investments in AI technologies and ensure that AI technologies are trusted.

The advent of AI is raising new issues and questions that must be carefully considered and addressed in order to support AI innovation and preserve timeless values such as respect for individual autonomy and privacy.

As AI is still at a nascent stage of development, an open dialogue between government, business, civil society and academic researchers is essential to shaping the continued development of the technology and realizing its potential benefits.

We believe that policy discussions should aim to promote broad development and deployment of AI across different sectors and continued AI innovation, encouraging outcomes that are aligned with the vision of human-centered AI.

We believe policymakers should:

Continue to convene broad dialogues among government, business, researchers, civil society and other interested stakeholders on how AI can be shaped to maximize its potential and mitigate its risks, including adoption of practical guiding principles to encourage development of human-centered AI

Stimulate the development and deployment of AI across all sectors and businesses of all sizes, including application of AI to address public and societal challenges, such as empowering underserved communities and those with disabilities, and adoption of AI in the public sector

Develop privacy laws with a view toward enabling the benefits of AI while preserving privacy

Invest in skills development and training initiatives for people at all stages of the job continuum

Encourage sharing of best practices in development and deployment of human-centered AI, through industry-led organizations such as Partnership on AI (PAI)

Fund short- and long-term multidisciplinary research and development of human-centered AI technologies and ways to use AI to provide insights into its potential socioeconomic impact

Develop shared public data sets and environment for AI training and testing to enable broader experimentation with AI and comparisons of alternative solutions to address ethical concerns

Evidence and further reading:

McKinsey Quarterly: An Executive's Guide to Machine Learning

Stanford University: Artificial Intelligence and Life in 2030: One Hundred Year Study on Artificial Intelligence: Report of the 2015-2016 Study Panel

Slate: The Partnership of the Future

The New York Times: The Race Is On to Control Artificial Intelligence, and Tech's Future

The AI: How Microsoft used AI to help crack down on tech support scams worldwide

Microsoft Green: Protecting Biodiversity with Artificial Intelligence

For links to these and other resources, please visit: http://www.microsoft.com/cloudforgood

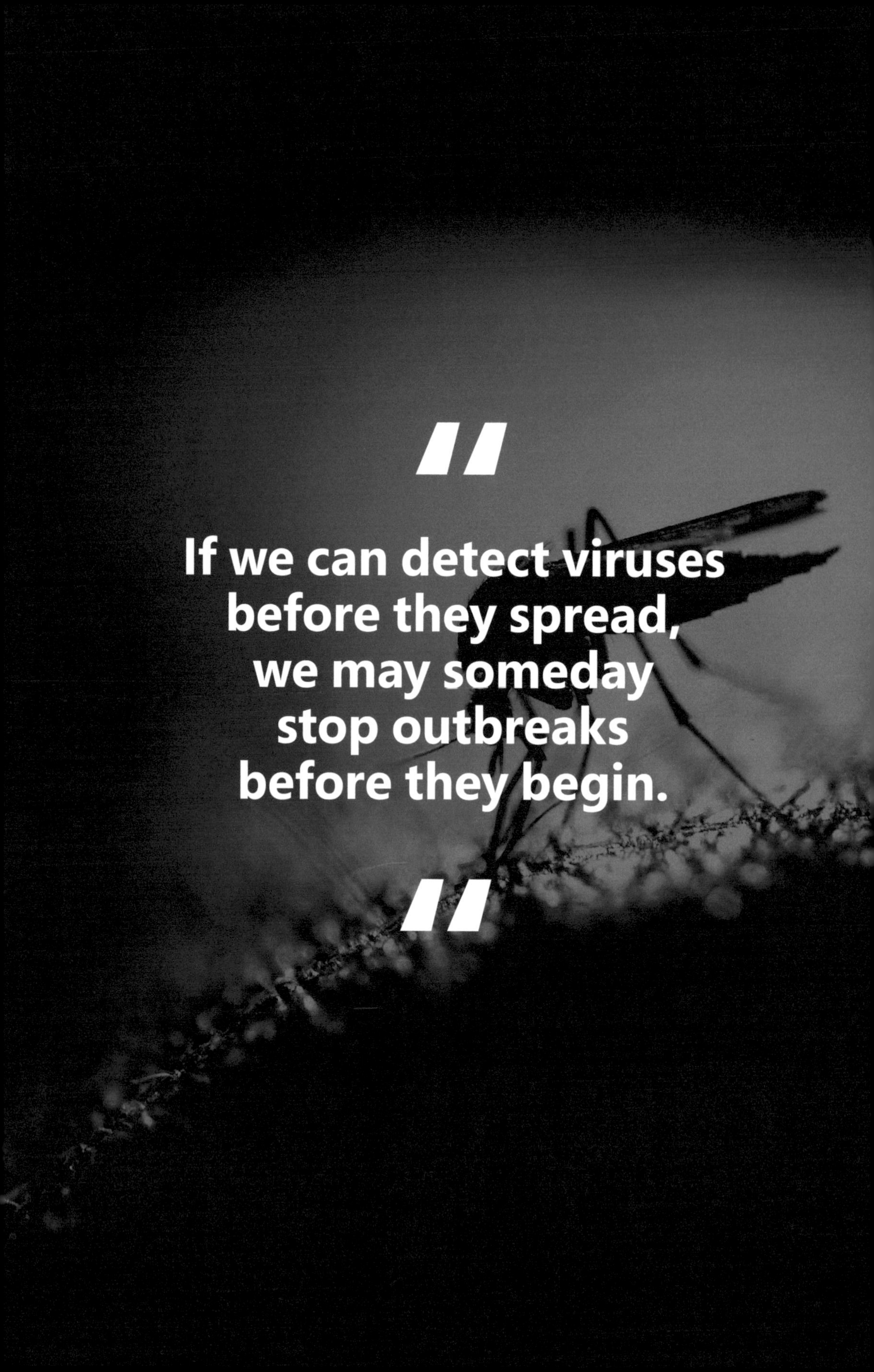

"

If we can detect viruses before they spread, we may someday stop outbreaks before they begin.

"

Project Premonition:

Using AI to track and stop outbreaks

Challenge

Each year, millions of people die from infectious diseases. And with 60 percent to 75 percent of emerging infectious diseases originating in animals, it's difficult to identify how and when a new disease occurs — or where it's spreading.

What if humans could prevent diseases such as the Ebola virus and avian flu before they become epidemics? That's the goal of Microsoft Research, which is combining AI and machine learning with drones, gene sequencing and cloud computing to pinpoint how, when and where infectious diseases occur.

Technology solution

Mosquitoes are the ultimate field biologists, taking blood samples from every animal they bite. By analyzing their DNA, researchers can identify both the animals that carry infectious diseases and the viruses they're carrying. The problem is that mosquito populations change with the weather, so finding where they hide can be difficult. Moreover, a single sample can contain 100 million pieces of DNA, making the process of accurately detecting pathogens slow and cumbersome.

Using AI to track and stop outbreaks

To identify emerging infectious diseases before they turn into outbreaks, Microsoft Research has launched Project Premonition. It uses Microsoft drones to find mosquito breeding grounds, robotic traps to gather specimens, and cloud-scale genomics powered by machine learning to search the DNA for pathogens.

Outcome

Using drones and robotic traps, researchers can more easily find and catch mosquitoes. And with technologies such as AI, machine learning and cloud computing, they can rapidly analyze hundreds of gigabytes of gene sequencing data. Thanks to Project Premonition, accurately detecting pathogens in a sample now takes fewer than 12 hours — versus the 30 days previously required.

Says Microsoft Researcher Ethan Jackson: "If we can detect viruses before they spread, we may someday stop outbreaks before they begin."

Microsoft Research Hardware Lab
Microsoft
Project Premonition
Microsoft

Policy section

An inclusive cloud

Policy recommendation

Providing affordable connectivity everywhere

The opportunity

The internet offers many social, economic and educational benefits for those with access. Currently, however, there are only about 3.4 billion people with an internet connection — slightly less than half the world's population.

Today, governments everywhere recognize that for all citizens to take full advantage of the benefits and opportunities promised by a new generation of innovative cloud-based technologies like AI, data analytics and the Internet of Things (IoT), affordable broadband internet access is essential. There is additional urgency for many countries because affordable internet access is a prerequisite for achieving the Sustainable Development Goals (SDGs) adopted by the United Nations in 2015.[22]

The challenge

As cloud-based services become increasingly central to our day-to-day lives, the need for robust, ubiquitous and affordable broadband connectivity becomes all the more critical. Significant gaps still exist across and even within countries.[23] For example, internet penetration is close to 100 percent in Korea, Qatar and Saudi Arabia, but below 2 percent in a number of sub-Saharan African nations.[24] Without specific steps to increase access, just 16 percent of people in the world's poorest countries and only 53 percent of the total global population will be connected to the internet in 2020. At this rate, universal internet access in low-income nations won't be achieved until 2042, a dozen years after the target date called for in the SDGs.[25]

Furthermore, the burgeoning Internet of Things (IoT) is creating even greater demand on internet access. Cisco's Visual Networking Index (VNI) forecasts that the number of connected devices will increase from 17 billion in 2016 to 27 billion in 2021. A majority of these will require some form of low-cost wireless connectivity to the cloud via the internet.

Policy recommendations

Successfully closing the gap for those unconnected will take innovation in policy, technology and business models. The size of the gap makes clear that we cannot close it simply through "business as usual." Certain policies can even create obstacles to investment by favoring a particular technology, industry sector or incumbent. Regulations that prescribe specific business models that must be used may in fact result in no competition, poor service and high prices. The pace of change is accelcrating, and agility is necessary for today's policymakers. The new challenge is to create the right enabling environment that will result in sustainable growth. To achieve this, promotion of the following is required.

Policy innovation: Reform of laws and regulations is needed to spur programs that expand internet access. Initiatives such as the Alliance for Affordable Internet, the U.N. Broadband Commission for Sustainable Development, and the 1 World Connected project play an important role in highlighting examples of policy innovation around the world and helping policymakers understand why some countries are succeeding in making internet access more affordable. Innovation may involve stimulating competition, removing financial barriers, altering tax policy, migrating government services and more. Creating open and competitive broadband markets in their own countries must be a core policy objective, and that means that policies to remove obstacles that limit opportunities to provide access are vitally important.

Technology innovation: Technology is changing rapidly — and nowhere is that more evident than in wireless communications. Our ability to harness spectrum for broadband has advanced exponentially over the past decade, with some wireless technologies able to provide gigabit-per-second connectivity. But access to spectrum often remains hindered by a regulatory model first developed over a century ago and optimized around exclusive use licenses protected by swaths of unused spectrum as a buffer from potential interference. Although this model is appropriate for some services, today's technologies allow for sharing of spectrum, resulting in greater reuse of spectrum at lower costs and improving access for all. Today, more than half of the world's internet traffic moves via unlicensed and shared spectrum. Policymakers should work to accelerate time to deployment of these new technologies by review and revising their approach to spectrum.

Spectrum management: Policymakers should accelerate efforts to open up new low-, mid- and high-frequency bands for both unlicensed and licensed use. TV White Space is one example of technology innovation in low-band spectrum. Controlled by databases, unlicensed access to TV White Space is delivering affordable internet connectivity today in several countries where regulators have allowed it — all without causing harmful interference to any protected service. Unlicensed spectrum is dramatically lower in cost than licensed spectrum because there are no auction or licensing fees that must be accounted for in an operator's business model, enabling more rapid and lower-cost deployment scenarios. Since withholding spectrum from deployment results in artificial scarcity, it tends to also cause higher prices for access. Policymakers should also ensure that any spectrum licensed on an exclusive basis is actually deployed via "use it or lose it" policies. Spectrum regulators should also adopt policies to facilitate the sharing of underutilized spectrum. In the long run, enabling dynamic access to shared spectrum resources is key to making spectrum an abundant resource, as the digital fuel to the digital economy.

Business model innovation: Today's telecommunications landscape is very different from that of the past, when national monopoly carriers delivered circuit-switched voice services with treaty-based international connections. Today's environment includes numerous public and private networks interconnected through a wide range of commercial agreements. Today's internet network of networks is the enabler for the digital transformation of business, government and leisure. Where the policy environment invites experimentation with new business models and partnerships, amazing transformations occur. Consider, for example, the impact of the availability of mobile money in Africa. Policymakers should be on the lookout for policy and regulation that inhibits this kind of innovation.

For example, when it comes to financing, many governments have imposed restrictions on foreign direct investments in telecommunications, mobile and broadband infrastructure, and have enacted other investment policies that effectively put barriers in the way of entrepreneurs willing to enter the market. Policies that encourage public-private partnerships and recognize the structuring needs of funding institutions are needed to facilitate access to capital.

In addition, in many countries, broadband access is taxed as a luxury good. This is counterproductive because it simultaneously reduces investment in infrastructure and increases the cost of access, as well as misunderstanding the importance of broadband access for utilizing everyday services. Policies like this contribute to the widening of economic and social divides. Policymakers should adopt tax policies that stimulate rapid investment in and adoption of connectivity solutions and that optimize taxation regimes to achieve connectivity goals.

Policymakers should, themselves, be innovative and adopt telecommunications policies that minimize unnecessary regulation of services and stimulate competition, while supporting the underlying goals of access and adoption. For example, reforming universal service funds to incorporate funding for broadband; ensuring net neutrality, which stimulates innovation in services; and reducing the regulatory burden on operators are all generally helpful. At the same time, the independence of regulators and freedom from regulatory capture are critical to stimulating investment.

Finally, there are a variety of approaches that can stimulate accelerated investment and deployment, such as "dig once" policies ensuring that new highway and rail infrastructure projects include conduits for fiber. In addition, infrastructure-sharing such as shared transmission towers can avoid wasteful duplication. Demand-side stimulation through deployment of online government services and the development of locally relevant content and services can also help.

Evidence and further reading:

Microsoft On the Issues: A rural broadband strategy: Connecting rural America to new opportunities

World Economic Forum: Internet for All: A Framework for Accelerating Internet Access and Adoption

WSJ: Rural America is stranded in the dial up age

Rural broadband progress report (2016)

The Economist: TV dinners: unused tv spectrums

Microsoft On the Issues: Using TV White Space technology in Puerto Rico and the US Virgin Islands

For links to these and other resources, please visit: http://www.microsoft.com/cloudforgood

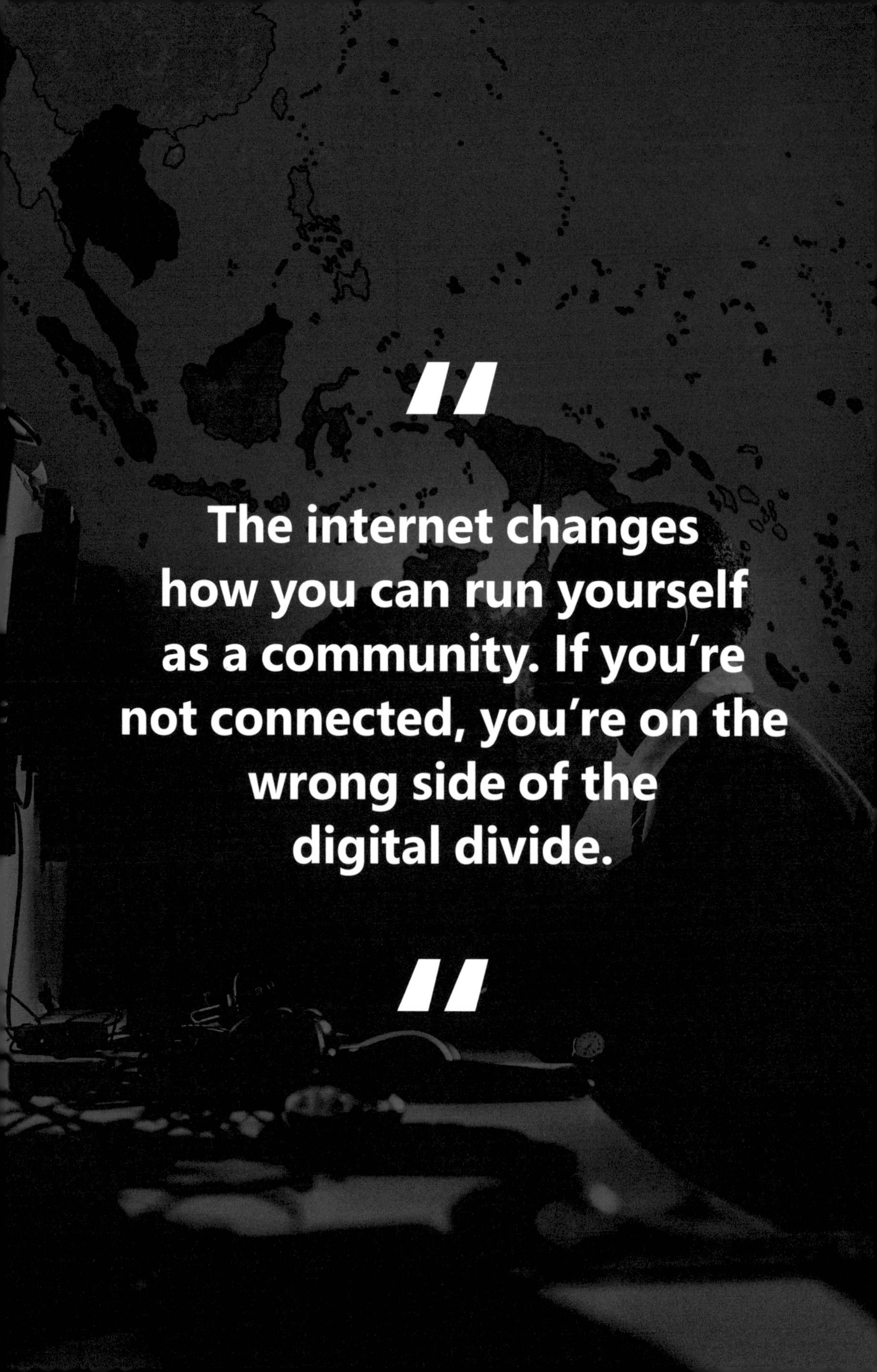

“

The internet changes how you can run yourself as a community. If you’re not connected, you’re on the wrong side of the digital divide.

”

Mawingu

Empowering Kenya and the world with high-speed, low-cost internet

Challenge

If cloud computing is the fundamental enabler of the next great technology revolution, then internet access is the essential prerequisite for reaping the benefits and opportunities that the cloud will unleash. For 4 billion people around the world, this is not an issue. But fairness and equity depend on ensuring that the 3.5 billion people who remain on the wrong side of the digital divide get access, too.

Nanyuki, Kenya—a town 125 miles north of Nairobi, where sending an email used to require a five-hour walk to the nearest wireless hot spot—offers hope that affordable access to the internet for all is achievable.

Technology Solution

Mawingu (which means "cloud" in Swahili), uses technology that takes advantage of underutilized television broadcast spectrum known as TV white spaces to provide low-cost internet access to residents of Nanyuki and the surrounding countryside.

Outcome

In its first three years, Mawingu has had a dramatic impact. By connecting county government offices, the county library, a Red Cross office, and a medical clinic to the outside world via the internet, Mawingu has improved access to public services and healthcare. At Gakawa Secondary School, one of five schools now connected to the internet, students have seen their scores improve on every single subject on the Kenya National Exam.

Mawingu also provides the infrastructure for new businesses such as Solar Cyber, an internet café that offers unlimited internet access for 3 U.S. dollars a month. Run out of an old 20-foot shipping container, Solar Cyber is used by farmers who check market prices for their crops, students who do homework, and young entrepreneurs who connect with customers around the world.

The success of Mawingu has attracted international funding to take the technology to a national scale. In September 2016, the Overseas Private Investment Corporation (OPIC), the U.S. government's development finance institution, committed to financing a 4.1 million U.S. dollar loan to help Mawingu provide affordable, solar-powered wireless internet to communities throughout rural Kenya. Research conducted by the World Bank a few years ago suggested that a 10 percent increase in internet access in Kenya would translate to a 1.38 percent jump in GDP.

As cloud computing drives a new generation of technology-driven innovations, the potential for growth could be even greater. This means the impact of TV white spaces in Nanyuki and other places like it where internet access has remained out of reach will truly be transformative.

Policy recommendation

Preparing people for the new world of work

The opportunity

As the types of jobs that constitute a modern economy continue to evolve, so do the skills required to do them. We need to ensure our systems for preparing, educating, training and retraining the current and future workforce keep up with the pace of digital transformation, ensuring that our citizens are well placed to benefit from the opportunities this transformation provides.

Not only are new skills in demand today, but there is a growing recognition that people will need to engage in regular reskilling throughout their working lives, including both technical skills and professional or "soft" skills. A person's ability to learn new things, collaborate, communicate and adapt to changing environments may in fact become the most important set of skills for long-term employability. Together, the increased needs from employers and individuals for the right training at the right time create opportunities for technology.

The challenge

For companies to thrive in a digital economy, the skills of employees must keep pace with advances in technology. When a mismatch occurs, the outcome can be devastating. In the manufacturing sector in the United States, as many as 2 million jobs could go unfilled during the next decade because of a shortage of people with the right technical skills.[26] In Europe, a 2013 survey found that skills shortages caused major business problems for a third of EU employers.[27] And in China, McKinsey estimates that demand for skilled labor could outstrip supply by 24 million people by 2020.[28] Shortages like these pose serious competitive issues for companies and threaten the long-term economic health of countries around the world. More than that, they threaten to widen the income gap between those who have the skills to succeed in the 21st century and those who do not.

Failing to address this mismatch between the business need and individual skills will leave many people facing an uncertain future — in particular, groups who are already at a disadvantage: women, young people, and those in rural and underserved communities. Closing this divide is an important factor in addressing income inequality and one of the most important actions governments can take to strengthen their economies. Recognizing this, the United Nations has established several related targets as part of the Sustainable Development Goals, including 4.b, which states: "By 2020, substantially expand globally the number of scholarships available to developing countries ... for enrolment in higher education, including vocational training and [ICT]."34

There is a great deal that employers can and must do to address the skills divide, from better sharing of data on their skills needs to new approaches to hiring to retraining and reskilling their own workforce. Yet their efforts alone will not be sufficient to reach the full need that exists. A CapGemini and MIT survey found that although 87 percent of companies feel that digital transformation is a competitive opportunity, only 46 percent were investing in developing the digital skills in their workforce needed to capitalize on these opportunities. If the situation is challenging for employed workers without ICT and technical skills, it is far worse for the unemployed and underskilled.

Policy recommendations

Digital literacy and ICT skills will increasingly be fundamental requirements for most jobs and for continued lifelong learning. According to a study of job postings by Burning Glass Technologies, 8 out of 10 middle-skills jobs require basic digital literacy as a prerequisite for employment.[29] Primary, secondary and postsecondary schools as well as worker retraining programs must offer technology education and information and communication technology skills training that is current, relevant and aligned with workforce demands. Fortunately, there are a range of innovative and low-cost approaches to help individuals become digitally literate. For its part, Microsoft has created and made available a wide range of curriculum, content and programs to support the needs of all learners across the digital skills spectrum — from foundational digital literacy to computer science education. Also, new computing devices and services — often enabled by the cloud — hold tremendous promise for one-to-one learning programs, offering richer, more personalized learning environments. Digital literacy can also be developed by integrating computing devices, software and online services into instruction for other subjects, which can help familiarize students with information and communication technology and cloud computing without them even realizing it.

Computer science and other science, technology, engineering and mathematics (STEM)-related skills are in demand and will require greater educational pathways throughout the education system. Policymakers must take steps to strengthen and increase computer science, data science, cybersecurity and STEM education throughout the education system with a particular focus on under-represented student populations. It is widely acknowledged that there is a mismatch between today's skill base and the future skills that are required in a data-driven and AI-enabled economy. Issues exist all along a training continuum that starts with primary and secondary education, extends to those entering the workforce, and continues to those who need to be reskilled and upskilled. The basic school curriculum in many countries still does not include exposure to computer-science education, which helps develop a learner's logic, problem-solving and creative reasoning — skills that are core to adaptability in today's work environment. In the United States alone, the Bureau of Labor Statistics estimates there will be more than 1 million American jobs created by 2020 that require a bachelor's degree or higher in a STEM field.

Promote entrepreneurial skills. Policymakers increasingly see entrepreneurs and small-business owners as essential to job creation and economic growth. Governments can promote entrepreneurship and small-business creation by partnering with companies and nonprofits to provide young people with the technology, skills and connections needed to launch businesses and create jobs for others. Building their businesses on cloud computing platforms allows them to pay only for the computing power their business needs and easily scale up as it grows.

The current workforce will need access to high-quality worker retraining programs and ongoing education. Governments should seek to meet the needs of people at all stages of the workforce continuum — students entering the workforce, unemployed and underemployed workers, and employed workers who need help gaining new skills to ensure their long-term employability. Governments should also think broadly about what training to offer and how to make it widely accessible. A first step is to identify the skills that are most in demand — a task that the IT industry is well-placed to assist with. With that knowledge, governments can develop and deliver high-quality workforce retraining programs or provide incentives and financial resources for private and nonprofit organizations to do so. Responsibility for identifying and addressing retraining needs shouldn't fall solely on governments. The private sector and educators also have an essential role to play. Education providers don't always offer training in the skills that employers are looking for; goals for educational attainment should also include outcomes related to employment, skills and advancement. The private sector has the real-world experience and insights to identify skills shortages and drive educational best practices. They are also essential partners in educational delivery. Solutions should consider blended learn-and-earn models such as registered apprenticeship and other models to combine secondary and post-secondary education and work in one experience.

Improve access to online services. The availability of online services in remote and underserved communities can be instrumental in expanding the quality and accessibility of education, training and broader civic engagement. The World Bank found that across 12 African countries, 9 percent of people with mobile phones or an internet connection use them to access formal education services every day, and 33 percent use the internet at least occasionally to find free education content.[30]

Encourage innovation and collaboration between the private, nonprofit and public sectors. Governments can also encourage entrepreneurship through programs that help people start new businesses. Programs that offer startups and entrepreneurs easy and affordable access to software, marketing support and visibility will help foster business success. Use technology and data to build a more dynamic skills-based labor marketplace that aligns job seekers to training and education providers to employers. Job seekers will better understand what skills are needed and where they can find the appropriate training and education, and employers can find the skilled talent they need. They will also enable education and workforce systems to access a variety of choices to achieve lasting career success, help employers find the skilled talent they need to grow, and make it possible for educators to train people with the skills required to compete in today's economy.

Evidence and further reading:

Microsoft On the Issues: Extending Microsoft's support for veterans entering the workforce

World Economic Forum: New Vision for Education: Unlocking the Potential of Technology

United Nations Resolution: Transforming our world: the 2030 Agenda for Sustainable Development

Deloitte Report: The skills gap in US manufacturing: 2015-2025 outlook

McKinsey & Company Report: Tackling Youth Unemployment

Malaysian Government Announcement: Creating a Nation of Digital Makers Key to Malaysia's Future Successes

Pew Research: Digital Readiness Gaps

For links to these and other resources, please visit: http://www.microsoft.com/cloudforgood

Policy recommendation

Including people with disabilities

The opportunity

Cloud computing offers significant benefits for people of all ages and abilities. It can empower people with visual, learning, age-related, mobility, hearing and speech disabilities to learn more effectively, engage and collaborate with others more easily, and express themselves more clearly. Cloud-based technologies such as natural language processing, AI and machine learning can deliver assistive technologies such as audio captioning and speech and image recognition. And because the cloud is a repository for custom settings, people can access information and services formatted to meet their preferences wherever they go, on almost any device.

Computers and the internet have also created opportunities for people with disabilities to pursue an education, participate in the workforce, access government services and engage in society. But a recent World Bank study found that people with disabilities are still significantly more likely to be unemployed and experience poverty. Cloud computing can help change this. Accessible cloud-based technologies can ensure that people with disabilities have access to services and content in today's increasingly digital world. It can also power new services that provide greater independence and mobility for people with a wide range of disabilities. Many of these innovations will also help people remain productive and independent as they age.

The challenge

There are more than 1 billion people around the world with disabilities.[31] Often, they face significant barriers when it comes to employment, education, healthcare, transportation, access to technology and government services, and more. Barriers to education for children with disabilities are particularly harmful because they have a lifelong impact on employability and contribute to high poverty rates for adults with disabilities.

Including people with disabilities

Although cloud computing already offers significant accessibility opportunities, people with disabilities are often the last to gain access to the benefits of technology-based innovation. Many cities are using cloud technologies to create new urban experiences, yet most of these "smart cities" are not fully accessible to people with disabilities, resulting in a growing digital divide. In rural areas, lack of connectivity can limit access to cloud-based accessibility solutions.

Policy recommendations

For people with disabilities, cloud computing can provide capabilities that improve access to education, employment, government services and rich engagement in society. To ensure that the benefits and opportunities provided by cloud computing are available to people with disabilities, governments should consider the following steps:

Procure accessible technologies. When public agencies purchase and use accessible technologies, access to public information and workforce participation by people with disabilities increases. By incorporating accessibility criteria such as ETSI EN 301 549 into procurement policies and requesting detailed compliance statements from technology and solutions providers, governments can create economic incentives for businesses to invest in more innovative accessible products. But although the United States and many European governments require federal agencies to procure accessible technologies, just one-third of countries that have signed the Convention on the Rights of Persons with Disabilities have enacted similar policies. All governments should sign the Global Initiative for Inclusive Information and Communication Technologies (G3ict) charter and integrate accessibility into their procurement procedures.

Promote globally harmonized standards. The speed of innovation in the technology sector can outpace legislation meant to encourage accessibility and promote cloud services. Global standards such ETSI EN 301 549 and ISO/IEC 40500 (W3C Web Content Accessibility Guidelines, known as WCAG 2.0) are regularly updated and thus better option than laws enshrining particular features or services and excluding others. By adopting public procurement policies based on these standards and requiring accessible technology from their suppliers, governments can create incentives for businesses to develop innovative, accessible products. These incentives are most effective when they are based on global standards that create unified digital markets.

Make smart cities inclusive. As cities embrace the cloud to create new urban experiences, they should be inclusive and not leave people with disabilities behind. Governments can demonstrate digital leadership and make a strong commitment to inclusion for all by implementing ICT accessibility standards and procurement policies. As civic leaders work to bring connectivity to rural communities, these same strategies can help ensure that people with disabilities can enjoy the same benefits from cloud technologies as their urban neighbors.

Embrace innovation. When governments try to solve complex challenges with obsolete technologies, people with disabilities are often left behind. Governments should adopt policies that encourage civil society, city leaders and educators to work with the technology industry to create new solutions that are based on cloud technologies and encourage their use to deliver education and public services.

Make e-government services accessible. In addition to procuring accessible technologies, governments should require documents, presentations, web content, applications and software solutions to be created and maintained in a manner consistent with applicable accessibility guidelines. This ensures that people of all abilities stay informed, participate fully in civic life, and take advantage of public-sector benefits, opportunities and employment.

Empower teachers. Teachers need to understand the value of accessible technology in the classroom. Education authorities should promote awareness and provide skills training to help teachers understand how to use new technologies to improve learning in the classroom. Most technology tools that are used in schools — such as computers, tablets and PCs — have features and support personalized settings that can help students access material and collaborate with peers, vastly improving accessibility in education. Skills training should help teachers take advantage of these accessibility features. Enabling teachers to create learning materials that are accessible to students with disabilities is also essential. When teachers understand how to adjust curriculum and content to the unique needs of their students, schools can successfully serve the needs of students with disabilities.

Evidence and further reading:

UNESCO: UNESCO Model Policy for Inclusive ICTs in Education for Persons with Disabilities

ISO: ISO/IEC 40500: 2012 Standard: Information technology -- W3C Web Content Accessibility Guidelines (WCAG) 2.0

Global Initiative for Inclusive Information and Communication Technologies (G3ict) charter: Promoting Global Digital Inclusion through ICT Procurement Policies & Accessibility Standards

Australian government announcement: Federal government will adopt the European standard for the procurement of accessible ICT (EN 301 549)

Microsoft white paper: Accessibility and the Cloud

Microsoft white paper: Accessibility Requirements for Public Procurement of ICT in Europe

Microsoft On the Issues: Creating incentives for digital inclusion: Microsoft announces support for G3ict charter for accessible technology

Smart Cities for All: Tools to help Smart Cities worldwide include a focus on ICT accessibility and the digital inclusion of persons with disabilities and older persons.

For links to these and other resources, please visit: http://www.microsoft.com/cloudforgood

Spotlight

Soft law and the benefits of standards

Remember when we used to talk about internet time? The world seemed to be changing so fast we could barely keep up. Looking back, it all seems rather leisurely, doesn't it? Now, with the cloud and AI, we seem to have made the leap from internet time to quantum speed almost overnight.

The question we face now is how to build trust in the cloud-driven technologies such as AI and data analytics that are rapidly transforming our world. As the speed of technological change quickens, we must also find ways to ensure that trust in technology keeps pace. This is complicated by the global nature of technology, which stands in contrast to the region-by-region approach that governments take to creating the rules that regulate the development and use of technology.

Although industry, businesses, civil society and individuals all have a role to play in developing trust in technologies, regulations are also needed. But there are many regulatory approaches, ranging from "hard" legislation to more nimble, "softer" approaches, such as the use of standards. Much of what makes technology succeed in the modern world is the creation of international standards, industry codes and government certifications. Standards help develop trust through clear definitions, best practices, transparency and proof (such as certification).

Each industrial revolution has seen the rise of new technologies that needed to earn people's trust. Because they did, we travel on airplanes without giving it a second thought; flip on a light switch without worrying about burns; and walk through town assuming bricks won't fall from buildings and injure us. Our packages arrive safely from wherever they were mailed, passing over the air, across seas, and on roads and rails to reach us.

Standards are at the core of this network of trust. In the IT industry, from the early days of hardware specifications, through the open standards that made the web possible, to today's work on the future of technology, standards play a critical role in developing confidence and understanding. Where IT standards once focused mainly on plug compatibility, protocol specifications and document formats, they now include critical risk management, behavioral and definitional agreements that are the foundation of the rules of cloud computing, AI, and cross-border data flows. In the International Standards Organization and International Electrotechnical Commission today, more than 100 countries participate to establish definitions, practices and transparency for new technologies. National representatives bring input from stakeholders within their country, ensuring that local interests have a voice in building trust through standards.

These modern cloud standards play a key role in important conversations, practices and transparency related to cloud services. For example:

Creating standardized language. Trust starts when we develop a common language so that words are interpreted in an established and uniform way. Standards create clear definitions that all stakeholders can rely on. Policy development and customer understanding benefit from specific definitions of different attributes of data and even cloud services themselves, and the same will be true for the fundamentals of AI and machine learning.

Developing best practices and structured transparency. Cloud services create powerful processing capabilities that may be housed a long way from users — even across national borders. Standards help create clear guidance for practices in areas like cybersecurity, privacy and accessibility. And they provide a common baseline and understanding to communicate those practices to users and governments so that everyone understands the protections afforded and can make well-informed decisions about cloud services.

Empowering users. Data has become a key value driver for the way organizations operate and interact with customers. Standards that guide data governance and service level agreements can help organizations as they embrace the digital transformation.

Providing assurance and certification. Trust is reinforced through proof points, everything from certifications to audits to attestations and more. These provide additional ways that customers and governments gain assurance for the practices, processes and technologies that process and store a customer's personal information.

International standards play a crucial role in guiding the behavior of cloud providers and helping regulators, customers and civil society understand what cloud providers are doing and how. With broad representation in the international standards system, the concepts established in standards can serve as foundational elements for key areas of policy and trust in cloud computing, even in areas as important and challenging as security and privacy. At Microsoft we are committed to investing the time and resources needed to work with industry partners, NGOs and governments to find the right balance between laws, regulation and standards.

Soft law and
the benefits
of standards

Spotlight:

What is artificial intelligence?

Artificial intelligence (AI) was once the stuff of science fiction. However, it is now no longer a promise; it is happening now, enriching lives through personal assistants such as Cortana, Siri and Alexa, and improving businesses in every industry, including healthcare, education, manufacturing, agriculture and transportation. For example, AI is helping hospitals more easily detect errors by recognizing anomalies in best clinical practices and save thousands of patients per year; enabling better analyses of how people learn throughout their lives so more effective and personalized approaches to education can be developed; and driving the development of precision farming techniques that can improve agricultural yield while reducing environmental impact.

It's possible now because of the computing power and storage made possible by cloud computing. The cloud, combined with significant progress in techniques to analyze massive amounts of data and draw insights, can be used to enhance every decision and process. Together, these advances have enabled great leaps forward in AI technologies that are giving our digital devices and systems

— everything from mobile phones to automobiles, airplanes and computers — the ability to perceive, learn, reason and make recommendations.

Simply put, AI is computational intelligence — a tool that is valuable in helping humans complete tasks and make decisions in a quicker and more effective manner. To give a sense of the speed at which AI can complete tasks, Microsoft's AI system can translate the English-language version of Wikipedia into another major language in less than one-tenth of a second, or the time it takes to blink an eye. It has also learned the ability to process human speech to the same level as a human and provide for real-time translation of conversations.

AI's ability to learn and process huge amounts of data offers immense potential to enrich people's lives, creating new opportunities for everyone and enabling humanity to tackle some of society's biggest challenges. AI stands to deliver a huge improvement in productivity, potentially helping to grow global GDP by an additional 26 percent by 2030,[32] if deployed successfully.

Research into AI has been taking place for a number of decades. However, recent developments have quickened the pace of progress. Advances in the creation of "neural networks," machine learning systems that enable computers to get better over time at recognizing patterns and making recommendations, have been particularly significant. This has led to breakthroughs in foundational AI capabilities such as natural language comprehension and speech, facial and object recognition.

Microsoft has a rich history in research and development of AI technologies and has been contributing to the advancement of AI for 25 years. Our vision is straightforward: **We aim to amplify human ingenuity with intelligent technology by making AI capabilities available broadly.** By empowering every individual to realize their full potential and every organization to innovate, AI can enable new socioeconomic opportunities and make it possible to address some of today's greatest global challenges.

We do this by:

Driving AI innovation that extends individuals' and organizations' capabilities and makes them more productive.

Building powerful AI platforms that make innovation by all developers faster and more accessible.

Infusing AI into all of our products and services, including Office, Windows, Xbox and more.

Taking a principled approach to creating human-centered AI that instills trust and keeps you in control.

You can learn more by visiting www.microsoft.com/ai

Chapter 3

Digital transformation in action

What does digital transformation look like? What happens when smart, creative, forward-thinking women and men take advantage of the power of the cloud to collect and analyze information at a scale and depth that has never before been possible? What problems will be solved and what unmet needs will be fulfilled as innovators and dreamers connect fresh insights with new digital capabilities?

These are questions that will take years to answer. History tells us that the full impact of an industrial revolution typically takes years to unfold and that the most important innovations were almost never foreseen by those who created the underlying advances that made each successive industrial revolution possible.

It was, for example, nearly a century after James Watt perfected the steam engine that people even began to call what he helped unleash an industrial revolution. It's doubtful that Heinrich Hertz, Guglielmo Marconi, and the other scientists and inventors whose discoveries laid the groundwork for radio broadcasts could have imagined the world of mobile devices and wireless communications that most people take for granted today.

But there are already exciting and inspiring hints of what the future may hold as people begin to discover how to use cloud computing, advanced analytics, mobile devices, connected sensors, genomics, 3-D printing, geolocation and a host of other related emerging technologies to look at old problems in new ways and to envision capabilities that until now were impossible to imagine.

Today, people in every profession are using cloud computing to work more efficiently and more effectively, to serve their customers in new ways, and to find solutions to once unsolvable problems. In this section, we offer a quick snapshot of the impact that cloud computing is having on industries that drive economic and social progress in communities around the world. Some of these stories are about small changes that have big impacts. Others touch on more substantial organization wide innovations. But, regardless of the nature of the organization, the country or region in which they are happening, or the services they describe, each one of these stories points the way forward toward a time when the digital transformation will enable people to strive to fulfill the deeply human desire to make the world a better place in powerful new ways.

Government

Although governments are continually under pressure to deliver more and better services to citizens, businesses and communities, they face new constraints on their financial and human resources.

Some governments are struggling to respond to unprecedented urban growth as people migrate to cities in search of new and better opportunities for jobs, health and education. Others struggle to respond to aging and more diverse populations. These challenges are compounded by citizens' growing expectations for faster and more accessible engagement with governments at every level — expectations fueled by social media and the advent of more personal and efficient interactions and experiences that businesses offer to consumers through mobile devices.

So how can cloud computing help governments address the challenges they face and better meet their citizens' needs and expectations? Although the process is still in its earliest phases, government agencies are beginning to take advantage of cloud-based technologies to transform the delivery of services — and even the fundamental nature of their relationship with the people they serve. Forward-thinking agencies are exploring ways to move beyond manual processes, disparate systems and paper-based methods as they work to unlock a digital future that opens the door to new business and government models, new services and new citizen experiences. Here are just a few examples of how governments at the national, state and local levels are transforming their effectiveness and meeting the changing needs of their citizens.

The **Tax Authority of Mexico (SAT)** is re-engineering its technology infrastructure to better serve the 40 million taxpayers who are required by law to interact with the authority online. Using the Microsoft cloud, the SAT has improved online services for citizens and businesses while increasing tax control and collection, lowering processing times, and reducing tax evasion.

Using the scalable storage and compute capacity of the cloud, the tax authority has significantly reduced disruptions in service that used to occur during spikes in activity as the deadline for filing taxes approached. The updated system has also strengthened the perception that the SAT is accountable to citizens and businesses, which has raised confidence in the agency, improved usage rates for the online tax service, and led to a 10 percent increase in tax revenues.

"With our new online portal, we're successfully encouraging taxpayers to file their taxes voluntarily, which means increased revenue collection."

Juan Manuel Galarza
General Administrator of Communications and Information
Tax Authority of Mexico

In Auckland, New Zealand's largest urban center, **Auckland Transport** is using advanced technology to help residents move around the city more efficiently. Working with Microsoft Services to develop a world-class transportation system, Auckland Transport now offers a wider range of transportation options by adding bus, train and ferry routes, while encouraging people to bike and walk. This effort includes a cloud-based website that is visited more than 1 million times per week, and the **MyStreet app** that tracks passengers' favorite routes, notifies them of street repairs and suggests detours. Auckland Transport is also using predictive analytics to assess data collected from sensors and social media and help drivers find parking and alleviate traffic congestion.

"Our greatest obstacle as a transportation agency is planning for population growth. To accommodate the projections, we need to increase efficiency and capacity, while keeping costs down and maintaining a positive experience for residents."

Roger Jones
Chief Information Officer
Auckland Transport

The **state government of Indiana** in the United States is improving access to an expanding array of business services such as licensing, tax payment and unemployment benefits through a one-stop portal called INBiz that is now used by 100,000 business owners. Built on the Microsoft cloud-based identity management system, INBiz provides personalized dashboards that alert business owners about upcoming filing deadlines, and has made submitting and processing information much faster and easier. As a result, Indiana has been able to expand and enhance its citizen services and increase compliance with the state's business regulations — all while improving cost-efficiency.

Education

As technology transforms our economies and creates new opportunities, the importance of high-quality education and the need to teach and train people effectively has never been clearer. Data-driven cloud technologies offer significant opportunities to improve the quality of teaching. New tools like AI and machine learning allow us to better understand how people learn and master new skills so we can identify more effective ways to teach. Virtual and mixed reality devices allow for new approaches to experiential learning. The global nature of the cloud provides new opportunities to connect with people and information from around the world, offering access to nearly unlimited learning resources.

So far, the impact of cloud computing on education has mostly been seen in cost savings and improved efficiency. Examples include e-textbooks, which are cheaper and more scalable than traditional textbooks, and subscription-based applications that reduce the cost of software, hardware and IT staff. But productivity and efficiency gains are just the start. Increasingly, cloud computing is helping schools increase student engagement and improve access to high-quality personalized learning. Here are some examples of how the cloud is transforming education and improving learning outcomes.

In the United States, **YES Prep school system** — which runs 16 schools in underserved communities in Houston, Texas — needed an identity access and management solution that would give teachers single-sign-on access to all the applications they rely on to help their students succeed. Today, YES Prep is using Microsoft's cloud suite of products to manage identity, mobile devices and security for its 1,000 faculty and staff members and its 11,600 students. This has reduced costs, increased application access flexibility and scalability, and given users richer identity management features. But, more important, YES Prep estimates that the new identity management system and student profile dashboard has reduced the time teachers spend on planning each semester by as much as 100 hours — time they now use to provide more individual instruction, engage with parents and further their professional development.

"Teachers were previously spending between 50 and 80 hours each semester to create differentiated learning plans, plus another 15 to 20 hours a week to keep the plans current. This goes to zero with our new student profile dashboard."

Richard Charlesworth
Chief Information Officer
YES Prep

In Canada, the **English Montreal School Board (EMSB)** serves more than 35,000 children and adults at 87 different locations on the eastern half of the Island of Montreal, including 4,000 elementary and secondary students who have learning difficulties, behavioral issues, or physical or intellectual disabilities. The district has 60 professionals — including psychologists, speech-language pathologists, guidance counselors, occupational therapists, behavioral management specialists and special education consultants — who provide services to students, parents, teachers and school administrators. In the past, extensive use of paper records meant professionals who needed a student's files had to drive to the EMSB central office or call to have them sent over, delaying access to the information. Filling out forms manually introduced the possibility of errors and resulted in duplicated efforts, with the same information on different forms. In addition, because paper files contained confidential student information, there was a risk of a privacy breach. Now, using a document management system called KiDS that runs on Microsoft Azure, the chances that private information will be accessed inappropriately have been sharply reduced, and the ability to create and access information in digital form saves each EMSB's special education professionals around five hours per week.

"With KiDS, our professionals work more effectively and efficiently, and that allows us to make better use of existing funds ... because of these efficiencies, it will make it possible for our professionals to help more students."

Luigi Di Filippo
Director of IT Services
English Montreal School Board

Healthcare

We live in a period of unprecedented progress in the improvement of healthcare for citizens around the world. Thanks to a combination of remarkable medical advances and a strong focus on global health, during the past quarter century average life expectancy around the world has increased by more than six years, and maternal and childhood mortality has been cut in half. Despite that progress, healthcare systems around the world still struggle with spiraling healthcare costs, with access to care due to aging populations or lack of care providers or facilities in remote areas, and with generating better patient outcomes.[33] Put simply, most healthcare systems are spending significantly more,[34] but still not reaching all citizens or delivering the results we would like to see.

Today, however, we are on the cusp of a period of tremendous opportunity, with new cloud-based technologies enabling us to unlock previously unknowable insight from patient data. These technologies allow healthcare providers to better collaborate in providing care and empowering patients.

Unlocking insight

Unlocking new insights from data will transform healthcare. AI, or the use of "machine learning" techniques on large data sets to extract insight and knowledge, is rapidly being integrated into the development of new tools and systems in the healthcare sector. Examples include tools like those produced by **EpiMed Solutions** that monitor entire hospital systems to detect and reduce the incidence of hospital-acquired infections by over 20 percent.[35] Predictive analytics tools that monitor usage by customers at Fullerton Health are being used to detect fraud and identify better care pathways, reducing care costs by 60 percent for healthcare plan members with chronic conditions.[36]

AI is also being infused into the next generation of clinical decision support and surgical tools. **365mc** in Korea is using cloud-based machine learning to create a surgical tool that will guide the stroke motion for the operating physician and provide warnings before the physician makes a critical mistake.[37] And on the diagnostics side, **Volpara Solutions** employs machine learning in one of its solutions to give technicians information in real time that enables personalized, high-quality breast cancer screening based on objective measurements of breast density, compression and radiation dose.[38] Without this insight, variations in tissue density have historically led technicians to perform scans that resulted in suboptimal resolution, which could impact the ability to detect cancer.

The cloud is also beginning to have an impact on front-line treatment by helping clinicians access evidenced-based medical standards and practices. Cochrane, a not-for-profit organization that creates, publishes and maintains systematic reviews of healthcare interventions, is now using AI and machine learning capabilities, built on the cloud, to bring together the best evidence from thousands of research papers and clinical trials to inform the development of treatment guidelines and healthcare policies.

Until now, rigorous systematic reviews could take up to two years to complete. Using AI to help select studies to be included in systematic reviews, Cochrane is saving weeks of work, freeing up healthcare reviewers to focus on high-level analysis. This system is helping clinical assessment groups look at the latest medical research in specialties ranging from cardiology and dementia to public health issues such as obesity, healthy eating and exercise. And it is now being used to speed the development of health guidelines by the National Institute for Health and Care Excellence in the U.K., which help U.K. National Health Service (NHS) doctors determine appropriate treatment for more than 65 million people.

"What excites me most about this work is that it is about exploring where those boundaries lie between what the machine can do and what the human can do … so that we can then better direct human effort where it's most needed."

Anna Noel-Storr
Information Specialist
Cochrane

Recognizing that AI offers great potential to patients but also raises important ethical and access-related questions, Microsoft has undertaken work to ensure that the benefits of AI in health do not just accrue to the privileged few. Through its Democratizing AI in Health Initiative, Microsoft will work to ensure we have a vibrant discussion platform to work through issues related to the ethics and access to AI-infused technologies in healthcare.[39]

Collaboration without boundaries

Cloud-based technologies also underlie new tools that drive greater collaboration among healthcare providers, across individual healthcare facilities and subject-matter experts, and increasingly across geographic boundaries. One of the greatest challenges with imaging technologies is the workload placed on the few radiologists who may be in proximity to more remotely located imaging devices. **Soddo Christian Hospital** boasts the only CT scanner in the entire southern half of Ethiopia, creating a massive and ongoing reliance for scans and interpretation. Unfortunately, its lone part-time radiologist could not manage the full-time patient load. Eventually, the hospital entered into an agreement with the **University of Alabama (UAB)** to help with offsite reading of images, but had some initial challenges getting the images into the hands of the university radiologists an ocean away. **RadConnect,** an Azure-based medical image-sharing SaaS solution developed by **NucleusHealth,** provides that critical link, allowing the UAB's board certified radiologists to reduce the backlog and enabling more timely analysis of images, which ultimately saves lives.[40] Soddo Christian Hospital recently admitted a 4-month-old female patient suffering from persistent opisthotonus (spinal spasms and arching) with a history of seizures. Initial diagnosis included tetanus and meningitis, but treatment was not working. A CT scan showed a meningeal enhancement, but not an abscess. Through RadConnect and its partners at UAB, Soddo Christian Hospital isolated the issue and developed the treatment plan, including a very tricky lumbar puncture — all to great success

Telehealth, remote monitoring and patient empowerment

The cloud is also changing the way patients access care. The Nordics, for instance, have been great leaders in using technology to expand the care continuum out of the traditional medical clinic and into the communities and homes of people who cannot make it to see a doctor or nurse. Access to healthcare facilities can be a challenge for residents of remote areas of Finland, where the nearest hospital or medical center may be hundreds of miles away. But an innovative virtual hospital program, a joint project involving all five of Finland's university hospital districts, is making access much easier. **The Helsinki and Uusimaa Hospital District (HUS)** coordinates the Virtual Hospital program and led technical development of the platform, based on Microsoft cloud solutions, that makes it possible to improve patient access to quality care, reduce costs and enable healthcare providers to treat more patients in less time.[41]

Childhood blindness occurs in Armenia at three times the U.S. rate. In response, Dr. Thomas Lee, director of the Vision Center and associate professor at the USC Roski Eye Institute, embarked on a partnership with the **Armenian EyeCare Project.** The partnership's mission is to train Armenian surgeons to reverse the effects of retinopathy of prematurity (ROP), a degenerative eye disease most commonly seen in premature babies that leads to severe vision loss or complete blindness if left untreated. This aggressive disease is nearly 100 percent curable in the United States, but without the same training, resources and medical expertise available in Armenia, ROP continues to contribute to the country's high rate of infant blindness. Fast-forward to today and Dr. Lee's vision of the partnership has developed into something much more exciting, leveraging technologies not available eight years ago to help train Armenian surgeons remotely. This includes using, for example, Skype for Business to observe the complicated eye procedures and communicate with the Armenian doctors in real time — all from his own hospital 7,000 miles away.[42]

In Trinidad and Tobago, diabetes occurs in significant numbers of the population and can cause serious and costly complications during pregnancy. **The Diabetes Education Research and Prevention Institute (DERPi)** of Trinidad and Tobago decided to tackle the problem by bringing new and easy-to-use technology to the patient. It created a Hyperglycemia in Pregnancy Trinidad and Tobago (HiPTT) app running on Azure that facilitates a system of data logging, analysis, visualization, archiving and communications necessary to provide efficient support to pregnant women and identify diabetes risk early in pregnancy.[43]

Challenges to access don't just stem from barriers like distance or reduced mobility; social stigma and cultural differences can also create complications that can now be addressed via technology. According to the World Health Organization (WHO), there are 23 million people suffering from mental health conditions in Egypt alone. Yet local cultural values often cause people to downplay or ignore entirely these conditions, preventing patients from getting any treatment. Egypt-based **Shezlong** is filling this service gap by offering online cloud-based treatment from therapists, helping to overcome cultural barriers and social stigma typically associated with seeking and receiving treatment for mental illness.[44]

Life sciences: Data-sharing scenarios

Beyond traditional healthcare, we also see emerging uses of cloud computing platforms to enable processing of large data sets across various private- and public-sector organizations in the pharma and life sciences sectors. Clinical research and pharmaceutical markets demand a one-stop-shop platform to aggregate research studies, provide an easy-to-use search interface, and offer tools for select data analysis. Vivli, a Massachusetts-based nonprofit, has collaborated with BlueMetal to build a first-of-its-kind data-sharing platform for clinical trial research, powered by Microsoft Azure. Using this platform, public and private institutions worldwide will be able to optimize time, effort and financial resources spent on research, as well as unearth new insights that will lead to critical discoveries.[45]

Cloud computing is also proving pivotal in the growth of the utility of biobanks for research. A biobank is a repository that stores biological samples (usually human) for use in research. Over the past 20 years biobanks have become an important resource in medical research, giving researchers access to data representing a large number of people. **BC Platforms,** a world leader in genomic data management and analysis solutions, operates **BC|RQUEST,** a unique resource that enables integrated analytics of genomic and clinical data, facilitating browsing and analytics of genomic and clinical data that has been aggregated across numerous distinct biobanks.[46]

Shaping the policy landscape

Although the promise of new technology is immense, we also recognize that many of these new technologies stretch often outdated regulatory policy beyond its limits. By working with healthcare providers, policymakers and patient groups, we have helped encourage a dialogue that balances the benefits of new technologies with the need to ensure transparency and protection of patient health information. The European Cloud in Health Advisory Council meets twice per year to create calls to action and other guidance for regulators and policymakers, enabling them to update regulations to reap benefits of new technology without undermining patient privacy protections.[47]

Capabilities like these are an important reason that the expanded use of digital technologies, including the cloud, is an essential component of healthcare policy in the European Union, a key part of the Affordable Care Act in the United States, and a pillar of the World Health Organization's long-term approach to improving health around the world. Working together, we can empower better citizen health with technology.

Manufacturing and industry

As the digital transformation moves forward, manufacturers are developing new capabilities to drive competitiveness and growth. The cloud allows manufacturers to reimagine how they engage with customers, empower employees, optimize operations and reinvent business models. International Data Corporation (IDC)'s latest research shows that manufacturers are leading the worldwide charge to invest in big data and business analytics solutions[2] and are likely to double the percentage of their products that are connected in the next three years[3] — all paving the way toward new revenue streams that will extend the lifetime value of a customer. Already, organizations that are taking advantage of data, analytics and the cloud generate an average of $100 million more operating income each year than those that don't. Here are a few stories of companies that are using cloud-enabled technologies to transform how they develop, manufacture and market products.

At part of its ongoing quest to use technology to make people's lives easier, the **Volvo Car Group** is adopting the mixed-reality capabilities of Microsoft HoloLens to revolutionize the way employees market and sell vehicles. With HoloLens, Volvo sales associates can offer customers a detailed, immersive view of the cars they're interested in — everything from viewing and choosing features, colors and options, to stripping away the outer layers of the vehicle to learn more about the engine, transmission and other technical details, to seeing safety features at work.

The ability to visualize and interact with products in deeper ways is also transforming Volvo's development processes, helping designers and engineers continue the company's mission of building the safest cars in the world. And the company expects that HoloLens will deliver value in other ways, including helping employees improve operations on the vehicle production line.

"With HoloLens we have the freedom to create a bespoke experience which customers can steer themselves. Imagine using mixed reality to choose the type of car you want."

Björn Annwall
Senior Vice President
Marketing, Sales and Service at Volvo Cars

For medical and pharmaceutical organizations, refrigerators are an essential tool for storing medicine and sensitive materials. To better serve its customers, **Liebherr,** a leading manufacturer of industrial and consumer equipment, has teamed up with Microsoft to create a new smart refrigerator that has precise electronic controls for accurate temperatures and efficient cooling for optimum storage and energy use.

There are often legal requirements to ensure that medicines are stored at proper temperatures. Liebherr has also developed a new pharmaceutical refrigerator that is equipped with intelligent connectivity that monitors and analyzes critical performance data remotely, including the accuracy and stability of the fridge's temperature and the state of the compressor and door. It features a communication module based on Windows 10 IoT Core that collects data and sends it to the Microsoft Azure cloud platform, where Microsoft Azure Stream Analytics generates real-time insights on temperature and other data. With the predictive analytic capabilities of Azure Machine Learning, changes in performance that indicate future problems can be detected so that repairs can be made before there is a failure.

Daimler Trucks North America (DTNA), which makes some of the most recognizable commercial vehicles on the road today, is committed to helping its customers minimize the cost of ownership of its trucks and buses. To do that, DTNA needs both an ongoing relationship with fleet owners and managers and a way to stay connected to the vehicles themselves. To understand how vehicles are performing, DTNA built Detroit Connect, a pioneering data capture and analysis program. Built on the Microsoft Azure platform, Detroit Connect collects performance information from vehicles and stores it in Azure. This way, fleet managers and owner-operators can know within minutes if an engine or an after-treatment system in their fleet has experienced a fault event, understand the severity, and know when, where and how to fix the underlying issue. Today, more than 200,000 trucks use Data Connect to maximize uptime and extend vehicle life.

"We know we can drastically improve the reliability and fuel economy of a vehicle through data gathering and analysis. For a fleet with 10,000 trucks, saving even 1 percent of fuel can translate into a tremendous cost savings."

Matt Pfaffenbach
Director of Connectivity
Daimler Trucks North America

Public safety

As has happened whenever disruptive innovations emerge, the development of the cloud and other digital technologies has been accompanied by the emergence of new risks. We have seen nation-states use digital tools to attack civilians and civilian infrastructure. Social media networks and other internet-based platforms have also been abused, sometimes in an attempt to influence elections and compromise democratic institutions. This corrodes confidence, undermines safety and threatens economic stability for people everywhere.

But there's another side to this story — the potential for a new generation of cloud-enabled innovations to drive significant improvements in public safety. Around the world, law enforcement, public safety and disaster response agencies are beginning to take advantage of cloud computing and advanced analytic capabilities to reduce costs and serve the public more effectively.

Innovations range from relatively simple steps to improve efficiency and reduce costs all the way to a complete reshaping of the way these agencies keep communities safe. The following are just a few examples of how law enforcement and disaster relief agencies are using the cloud to transform their work.

In the Netherlands, the **Dutch National Police Corps** is using social media to collect timely information about incidents so it can deploy resources more quickly and effectively. Using the social media monitoring tool **PublicSonar,** which is hosted in the Microsoft Azure Cloud, the agency can identify when a potentially dangerous situation is emerging and provide on-scene officers with information pulled from social media platforms to help them take preventive steps as quickly as possible. Thanks to PublicSonar, the Dutch National Police Corps now has, in effect, citizens of all ages and from all across the country working with the police force.

These new technologies are transforming how evidence is collected and used by law enforcement. **Genetec,** a company based in Canada, is helping agencies around the world gather and share video and other types of digital evidence. Managing digital evidence from diverse sources is complicated. Files must be captured, copied and converted to a standard format. In addition, there can be significant security and jurisdictional issues involved in sharing data between organizations. Genetec is addressing these challenges through solutions such as Genetec Security Center, which offers video surveillance, access control and license plate recognition tools in a single cloud-based platform that also integrates external business and security systems. Genetec Clearance gives organizations ranging from small private security teams to large government agencies access to automated case management tools.

The cloud is also playing an important role in disaster preparedness. In earthquake-prone Mexico, the government's earthquake early-warning system operates only in Mexico City. To expand the solution's geographic coverage, SkyAlert built a solution on the Microsoft Azure cloud platform that uses the government's sensor network to alert people outside Mexico City about an impending earthquake on mobile apps and desktop computers. Already, 3 million people and more than 40 organizations use SkyAlert, which can warn people about an impending earthquake up to two minutes before it strikes — 40 seconds faster than the government's system.

"People want to be informed about everything related to quakes. They know that having an alert that also provides some information about the event will help them make better decisions when one occurs."

Alvaro Velasco
Director of Marketing and Apps Development
SkyAlert

Small and midsize businesses

Recent research makes clear that small and midsize businesses play a fundamental role in job creation, economic growth and the health of local communities. According to the United Nations International Labour Organization and the Organisation for Economic Co-operation and Development, unregistered small and midsize companies make up 95 percent of the world's enterprises. And worldwide, businesses with fewer than 100 people generate more than 50 percent of net job creation.[48]

Cloud-enabled capabilities can improve the competitiveness of small businesses, which can significantly expand access to job opportunities and strengthen local economies. Already, many small organizations are using cloud computing to lower costs, improve productivity and efficiency, and enhance agility. According to a recent study conducted by Pb7 Research and the software company Exact, small businesses that have adopted cloud solutions have doubled profits and increased revenue by 25 percent.[49]

Now, small businesses are taking advantage of cloud computing to build thriving businesses by reaching and serving customers in ways that would never otherwise be possible. Here are two stories of businesses that are taking advantage of the digital transformation to reimagine how they operate.

Over the past two years, **CityBee** — the only car- and bike-sharing service Eastern Europe — has seen its fleet grow by a factor of 10. Because renting a car or bike requires several clicks in a mobile app or online, the company needs a seamless and reliable technical infrastructure. With demand growing and high-quality services essential, CityBee implemented a cloud solution based on Microsoft Azure that offers flexibility, reliability and interoperability with other platforms and systems. Improved flexibility means the company can now forecast an increase in demand and modify its infrastructure almost instantly.

In Argentina, **Garantizar** is enabling 14,000 small and medium-sized enterprises to meet their capital needs by helping them manage loans from financial institutions. A dynamic organization that believes the human factor is essential to day-to-day operations, Garantizar depends on a great deal of interaction between different sectors of the company to ensure accurate and quick responses to customer requests. To help maintain this interpersonal approach, Garantizar has embraced the cloud for productivity, in-house communications and internal interaction. Microsoft Office 365 and Yammer have been the catalyst for a new era at the company where more social interaction between people in the organization has inspired greater employee loyalty and increased agility in how it shares news and information about important events.

Farming and agriculture

There may be no area where the impact of cloud computing will be more important — or, for many people, more surprising — than farming and agriculture. As the world's population grows from 7.5 billion today to 10 billion people in 2050, the combination of an expanding global population, rising incomes and changing diets will require the world's farmers and fishermen to produce about 60 percent more food by 2050, on less land and using less water.47

But although it's easy to imagine how the cloud might transform information-based sectors such as healthcare, financial services and manufacturing, it can be a little more difficult to understand how the ability to collect, store and process large amounts of information in datacenters around the world will help the people who grow food, raise livestock and catch fish be more productive.

In truth, digital technology has already had a significant impact on farming and fishing around the world. Today's farmers are no strangers to technology. From helping them know how densely to plant crops on an angled slope to identifying and applying modern best practices, technology is playing a key role in increasing agricultural productivity. In developed nations, many farmers rely on a wide range of innovative new technologies for their day-to-day operations — everything from self-driving tractors that use GPS and satellite imagery to plant more efficiently to drones and sensors coupled with advanced software that enable them to make smarter decisions about when to irrigate and how much fertilizer to apply. The following is a small sample of the impact of digital transformation on one of the world's oldest industries.

Land O'Lakes is a $13 billion agricultural cooperative in the United States with more than 4,000 member-owners, including agriculture and dairy producers and retailers who operate thousands of locations across the United States, selling the agricultural output of some 300,000 farmers. Land O'Lakes employees, who are often in the field working alongside farmers, need instant access to information and their colleagues, anywhere and on any device.

"Technology adoption has enabled one of the greatest productivity stories of our generation and resulted in lifting millions out of hunger by reducing the cost of the food we eat."

Michael Macrie
Senior Vice President and CIO
Land O'Lakes Inc.

To create a more flexible, collaborative environment, the organization is using the Microsoft Office 365 suite of cloud-based services to make it easier for employees to work together by crowdsourcing ideas, and more efficient to access information through a self-service solution that connects them to knowledge from around the company. Now, when they are working with growers, they can pull up crop yield data and satellite imagery to show a farmer which valley is running low on nitrogen and where they will get greater output. This is just the beginning of Land O'Lakes' efforts to unleash the power of the digital transformation to grow more food, foster new ideas and talent, and preserve communities and the planet for future generations.

A virus known as Pacific oyster mortality syndrome (POMS) virus has been devastating to oyster farmers in Australia's Barilla Bay in the southern state of Tasmania. **Barilla Bay Oysters** is one producer that has suffered — in February 2017, 70 percent of its harvest was wiped out by POMS. Oysters are especially susceptible when flooding following heavy rain affects water quality in the estuaries where they grow. To protect people from eating oysters that could make them sick, oyster farmers are required to shut down oyster beds whenever there is a potential risk. The decisions are largely based on readings from rainfall gauges and 30 percent of closures turn out to be unwarranted. The economic impact can be huge — a single day of lost production during peak season can mean AU$120,000 in lost revenue.

To protect consumer health and preserve the economic viability of the local oyster industry, agriculture technology innovator **The Yield** has developed a new system that uses sensors, cloud computing and machine learning to measure everything from water salinity, temperature and depth to barometric pressure and sea tide height. That data is used to understand and predict the conditions that make a POMS outbreak likely. The solution also offers a portal for food safety regulators that can help them manage openings and closings and share information quickly with oyster famers so they can respond accordingly.

"We actually have a really clear purpose, which is how we're going to help feed the world without wrecking the planet."

Ros Harvey
Founder and CEO
The Yield

Financial services

From maintaining stable currencies to providing insurance products, managing systemic risk, allocating capital, enabling home ownership and administering retirement portfolios, financial services institutions provide the infrastructure that helps businesses and consumers reach their goals. But the confluence of technology advances, cultural shifts and regulatory change has brought the industry to a pivotal moment. Innovative companies are disrupting old ways of doing business by delivering cloud-based products and services that offer greater convenience and lower costs. In the face of these challenges, financial service institutions have a unique opportunity to redefine their role as trusted brokers in the new economy. Their business imperative is to innovate, demonstrate relevance and deliver value on a foundation of trust. Here are a few examples of institutions that are doing just that.

AXA Global P&C uses sophisticated actuarial studies and natural catastrophe modeling to manage reinsurance programs for the AXA Group, a global insurance provider based in Paris, France. To improve its catastrophe models, simulate multiple flood scenarios on-demand, and expand its models to other types of natural disasters such as windstorms, the actuarial team created a high-performance computing (HPC) solution based on the Microsoft Azure platform and Azure HPC Pack. The Azure solution not only met the requirement to scale on-demand, but it enabled actuaries to use an open source software environment and the programming language favored by researchers engaged in statistical and predictive analytics. This has enabled AXA Global P&C to realize some of its more far-reaching goals — to expand beyond flood simulation to include other events such as windstorms and earthquakes to help people understand natural disaster risks to better protect lives and property.

"We could never go back from the cloud. Choosing Azure was an essential step in our evolution. We wouldn't have been able to develop natural disaster models without it."

Simon Blaquière
Reinsurance Actuarial Manager
AXA Global P&C

With a growing presence in China, Southeast Asia and South Asia, Singapore-based **DBS Bank** strives to be the bank of choice for the ever-expanding Asia economic region. To achieve this mission, DBS is digitizing its core banking activities, guiding customers to embrace digital practices, and transforming the relationship between employees and technology to create a more productive workplace. The first bank in Singapore to adopt the Microsoft Office 365 cloud-based productivity solution in the workplace, DBS is improving personal and organizational productivity, and empowering employees to connect and engage with customers in new ways as they gain new insights that will ultimately help the bank transform its products and services. Over the past few years, DBS has made headway in creating a workforce that uses the cloud to make the customer experience simpler and more seamless, enhance collaboration, increase efficiency, and focus on delivering greater customer value.

One of South Africa's major financial institutions, **Nedbank** has operations in seven African countries. To achieve its ambitious expansion plans, Nedbank is working to simultaneously improve client services and lower costs. One way it is doing this is to focus on replicating the convenience of call center interactions in more cost-effective channels, particularly those that its clients prefer, such as WhatsApp, Facebook Messenger, Slack and other texting apps. To encourage clients to engage with the bank on these channels, Nedbank used the cloud-based Microsoft Bot Framework to create the Electronic Virtual Assistant (EVA), a virtual call center solution that can understand the context of clients' questions at 10 percent of the cost of live agents. EVA works so well that many clients believe they are interacting with live agents; it handles 80 percent of the inquiries it receives, freeing live agents to handle the most complicated customers. For Nedbank, the bot technology is just the first step in a digital transformation that is driving the bank's vision to be the most admired financial services provider in Africa.

"The shift to serving individual investors is a priority for us in the next 12 months. Bot technology plays a big role in that."

Steven Goodrich
Head of Technology
Nedbank Wealth at Nedbank

Nonprofit organizations

Whether it's fighting inequality, addressing poverty or protecting the planet, the work of nonprofits is vital. But there's a problem: Funding is flat and demand for services is growing. One way that nonprofits are meeting this challenge is to implement their own digital transformation. Modern nonprofits are achieving efficiency gains and rapidly scaling their operations by adopting the same technology platforms that have enabled the private sector to innovate and grow. By using cloud-enabled technologies to move from building and managing technology solutions themselves to taking advantage of subscription-based offerings, nonprofits are reducing capital expenditures on hardware and accessing a broad technical support and development ecosystem that offers higher levels of innovation. And they are freeing staff to focus on their organization's core mission. Here are a few examples of nonprofits that are using to technology to improve their ability to realize their mission to serve others.

Every minute counts when responding to war, natural disaster or an outbreak of a deadly disease. Because saving time means saving lives, **SOS Children's Villages International,** which works in 135 countries to care for children, was looking for new ways to respond to the overwhelming need of refugees from the civil war in Syria. Taking advantage of Microsoft's Tech4Good program — an initiative that provides donated or discounted cloud services to eligible nonprofits — SOS is now using the cloud to connect employees, other aid organizations and the victims of the civil war.

For example, Skype for Business enables SOS workers in the field to stay in close touch with psychologists and other specialists to help children work through trauma and grief. Child protection staff can store and access case documents using a team SharePoint folder so they can work together to solve problems and share what they've learned. And with nearly 5 million people having fled the violence in Syria, SOS has set up information and communications technology corners along refugee routes in Europe so people can research upcoming stretches on their journey, seek medical advice, prepare to enter a host country, and — perhaps most important — stay connected to loved ones.

Compassion International provides health, educational, social and spiritual services for 1.8 million children in 25 countries through sponsorships. Looking for ways to increase its impact, Compassion International examined its approach to IT and discovered significant opportunities to centralize its digital infrastructure. Today, Compassion International is using digital technology not only to serve more children — about 7 percent more each year — but also to improve the impact it is having on the children it already supports.

The impact of this digital transformation is impressive. The time it takes to connect children and sponsors has been reduced from months to a single day. Moving from using paper forms in the field to digital information delivered via mobile devices and Microsoft Azure has streamlined processes and is saving more than 300,000 hours of staff time per year. The time it takes for a child to communicate with a sponsor — accounting for local technologies, the speed of international mail and translations — has dropped from six months to a few weeks. The results don't just save time, they are transforming Compassion International's ability to achieve its mission to improve lives.

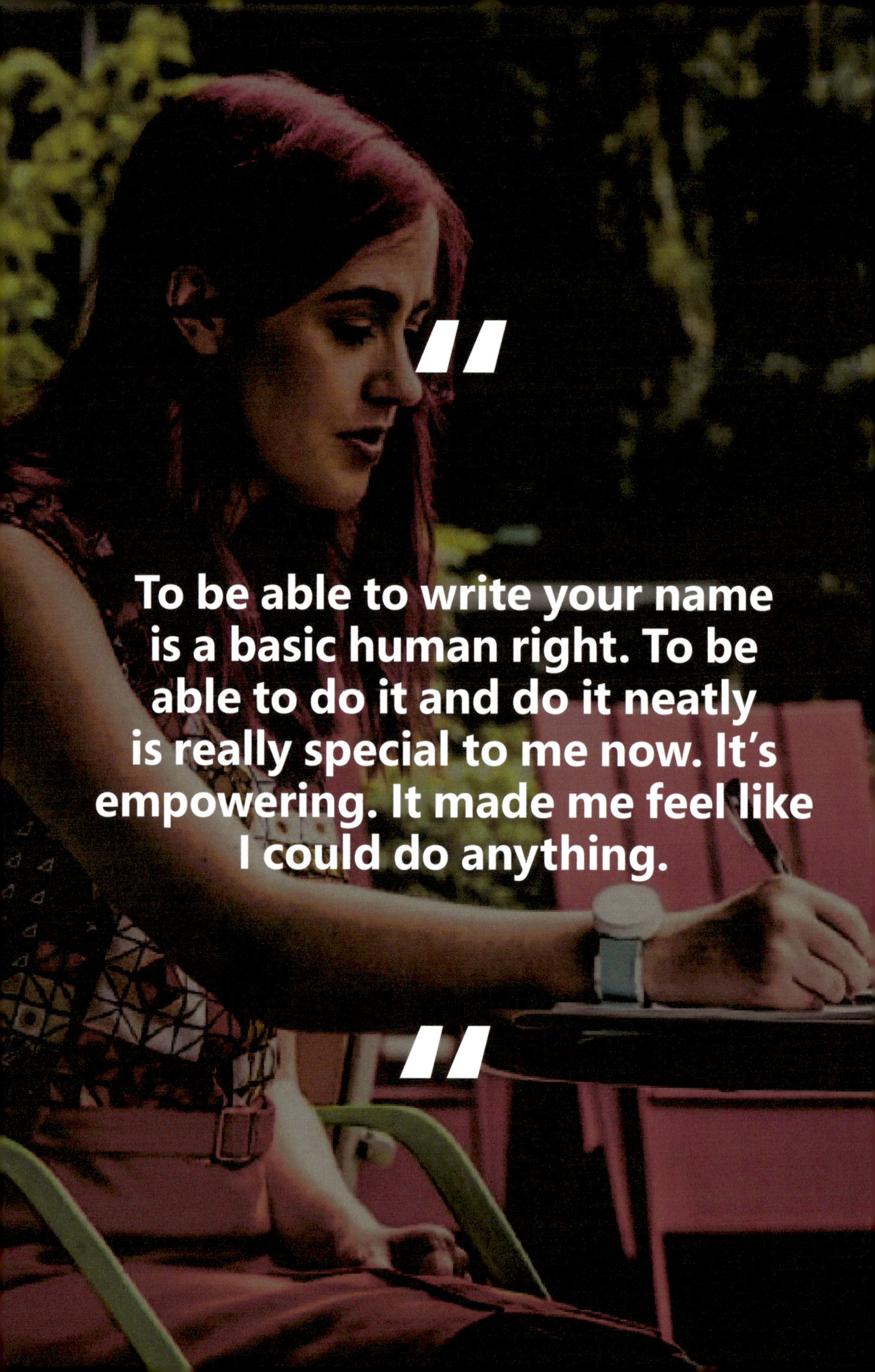
"
To be able to write your name is a basic human right. To be able to do it and do it neatly is really special to me now. It's empowering. It made me feel like I could do anything.
"

Project Emma:

How Emma was able to write and draw again

Challenge

Parkinson's disease is a chronic and progressive movement disorder that affects more than 10 million people worldwide. In people with Parkinson's, the brain fires extra signals to muscles, creating a chaotic internal feedback loop that causes muscles to panic and perform many movements at once. This creates tremors, which often deprive people of their ability to write legibly, type and draw.

Technology solution

Working with Emma Lawton, a London-based graphic designer diagnosed with Parkinson's in 2013, Microsoft researcher Haiyan Zhang researched the cause of tremors and designed a wearable device to reduce them. Guided by her passion to use technology for good, Zhang also looked to the Maker Movement's pillars of artistry, altruism and engineering.

She tested a special wristwatch, using tiny motors to cause vibrations that counteract the flood of messages between Lawton's brain and wrist. A custom app on Lawton's Windows 10 tablet controls vibration speed and frequency.

Outcome

How Emma was able to write and draw again

Wearing the Emma Watch, Lawton can write and draw again. Vibrations from the watch focus her brain on her right wrist and interrupt the extra signals — similar to injecting white noise into a feedback loop to disrupt it.

Now Zhang and her Microsoft Research colleagues are collaborating on a new initiative, Project Emma. They'll explore how sensors, AI, and machine learning can detect and monitor Parkinson's symptoms, including body rigidity, gait slowness, falling and tremors. The goal is to use AI and wearables to help people living with the disorder better manage their symptoms and medication intake — ultimately improving their quality of life.

"IT'S NOT GOING TO BE PERFECT... BUT, MY GOD, IT'S BETTER.

—"

"

This app, built here in Nepal by Nepali developers, can be the template. We can hand it off to the next government right at the beginning, so they can record where people live, where they come from, everything else, and store all of that information both on-premises and in the cloud.

"

Nepal earthquake:

UN uses cloud to revolutionize disaster relief

Challenge

In April 2015, a 7.8 magnitude earthquake rocked Nepal and affected the lives of approximately 8 million people. Nearly 9,000 people died, and some 600,000 homes and buildings were destroyed or severely damaged. With many of the survivors living in tents, there was tremendous urgency to begin reconstruction before the brutal winter set in.

Technology solution

The daunting task of rebuilding began with mapping where the original structures had stood. In the past, such records were maintained on paper, which can be difficult or impossible to obtain following a disaster of this magnitude. To expedite reconstruction, the Microsoft Innovation Center (MIC) Nepal — in collaboration with the United Nations Development Programme (UNDP) — built a mobile phone app that tracks and coordinates logistics, personnel and payments.

Built using Visual Studio and connected to Office 365, Power BI, SQL Server and Microsoft Azure, the app was available through the Microsoft Store to Windows Phones in the field. It offered a flexible,

UN uses cloud to revolutionize disaster relief

cost-effective solution for overseeing a complicated recovery and rebuilding operation in a remote area that otherwise would have required far more time, resources and money.

Outcome

The cloud-based app tapped into a mobile phone's GPS capabilities, allowing reconstruction crews to record precise coordinates and measurements for each building prior to demolition. The app also was used to manage daily cash payments to thousands of local workers, many of whom were clearing debris. That helped restart the country's economy and infused money into communities that needed it most.

The app from UNDP and Microsoft has the potential to revolutionize how the world responds to disasters. Storing records securely in the cloud can help rebuild census data, property records and land tenure rights — critical for recovery efforts in developing nations with rudimentary record-keeping systems. The app also can enable relief agencies to share valuable records created during early phases of a disaster to jump-start reconstruction.

"It's so huge," says Jamie McGoldrick, who oversees the UNDP in Nepal. "We can revolutionize the entire global response to disasters. With a cell phone, we can change the world."

"

Artificial intelligence is improving at an ever faster rate, and I'm really excited to see where we can take this. I think this is just the beginning.

"

Seeing AI:

How AI is helping blind and low-vision people navigate the world

Challenge

Blindness affects 36 million people worldwide, and another 217 million have moderate to severe vision impairment. For them, everyday tasks such as identifying products in a grocery store, reading restaurant menus and seeing who's around them when they walk down the street can be challenging.

Technology solution

Microsoft software engineer Saqib Shaikh understands this firsthand. He lost his eyesight at age 7 and began using talking computers at a school for the blind. Seeing the positive impact technology can have, he envisioned using computers to improve life for the low-vision community. "One of the things I've always dreamt of since I was at university was this idea of something that could tell you at any moment what's going on all around you," he says.

How AI is helping blind and low-vision people navigate the world

Shaikh and his colleagues have been exploring how AI can empower the blind and low-vision community to experience their surroundings more vividly. Using Microsoft Cognitive Services APIs and AI technologies such as machine learning, the engineers have built "Seeing AI," an app that reveals the visual world to blind and low-vision people. The technology can read text out loud, recognize people and their emotions, and even describe everyday scenes, such as a skateboarder performing a trick.

Outcome

With Seeing AI, blind and low-vision people can use their iPhone or Pivothead SMART eyeglasses camera to better navigate their surroundings. The free app can identify a product by its barcode and call out the name. It can read documents, including headings, paragraphs and lists, allowing users to skim through to find the text they need.

It can also recognize people based on their faces and provide a description of their appearance, including their gender and facial expression. For example, the app can help users identify who they're talking to by telling them, "I see two faces: 40-year-old man with a beard looking surprised and a 20-year-old woman looking happy." Likewise, if a low-vision person is sitting in the park and hears nearby movement, the app can tell them: "I think it's a young girl throwing an orange Frisbee."

Microsoft's goal is to get Seeing AI technology into the hands of as many blind and low-vision people as possible. "For me, it's about taking that far-off dream and building it one step at a time," Shaikh says. "Artificial intelligence is improving at an ever faster rate, and I'm really excited to see where we can take this. I think this is just the beginning."

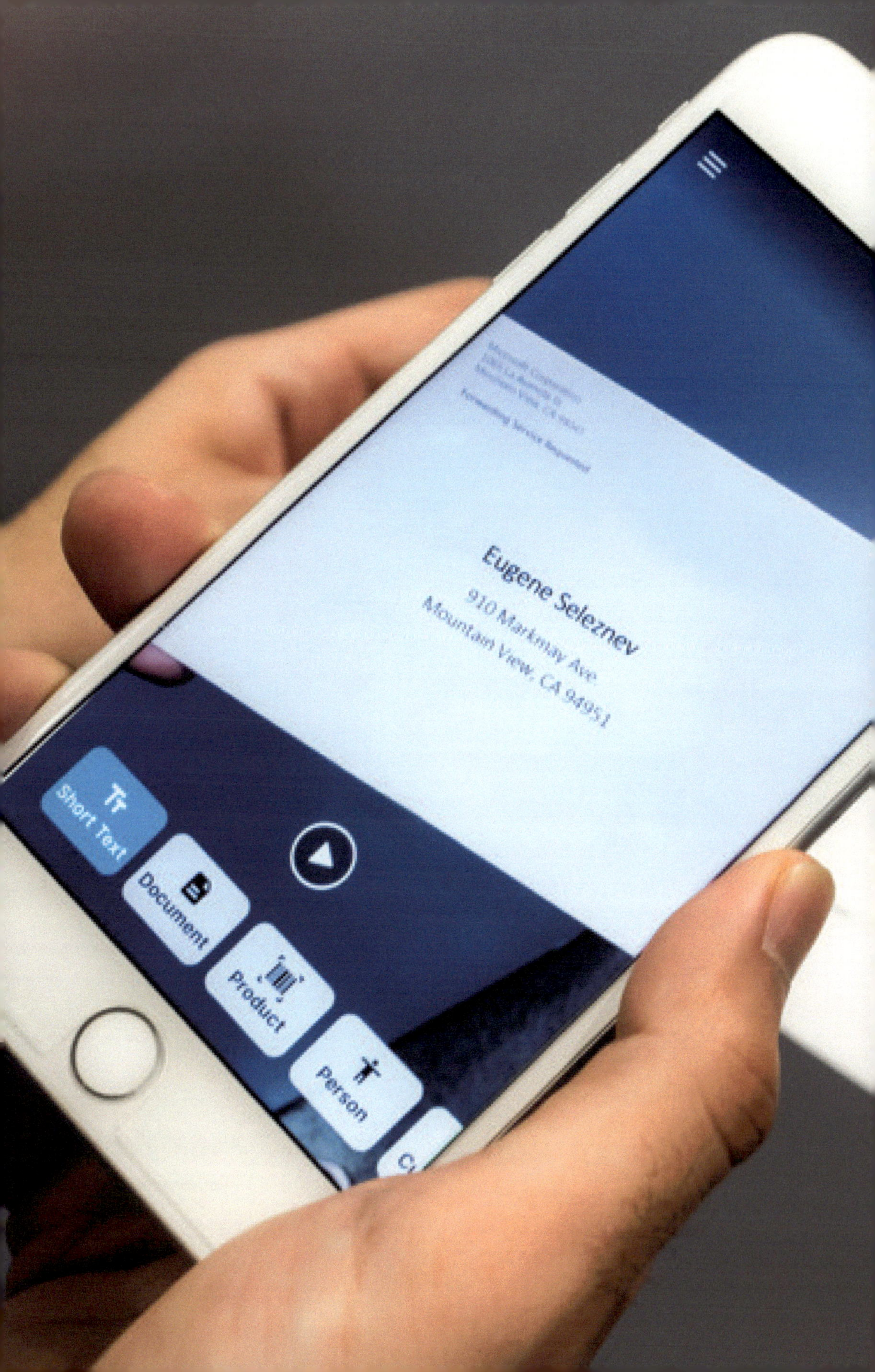
Eugene Seleznev
910 Markmay Ave
Mountain View, CA 94951
Short Text
Document
Product
Person

Spotlight

Welcome to the quantum age

The modern world has been shaped by computers. The exponential increase of computing power driven by Moore's Law, combined with the incredible expansion of data stored in the cloud and developments in machine learning, has led to the rise of AI, which will reshape our world again. And beyond AI and the Fourth Industrial Revolution, quantum computing is now emerging as another significant technology advance that has the potential to fundamentally change our world.

The computing power provided by quantum computing would enable us to make huge strides in understanding and addressing some of the major challenges that we face as a society. By improving our ability to process data about how cancers respond to treatments or how our environment is affected by human activity, this vastly increased computing power will serve as a powerful new tool. However, quantum computers will do more than just improve our ability to run calculations.

Because quantum computers are built as quantum systems, they will allow us to model other quantum systems virtually without having to construct them in the real world. This would allow us to easily model the types of experiments currently being conducted in high-energy particle accelerators like the Large Hadron Collider at CERN, or the types of chemical reactions only currently observable in laboratories, all on a computer. This ability could help us make dramatic progress in, for example, developing a catalyst to remove carbon from the atmosphere, or creating new enzymes to speed food production, or targeting medicine in a highly precise way to treat diseases.

Quantum computing takes a giant leap forward from today's technology — one that could transform the economic, industrial, academic and societal landscape. It makes use of the way the universe works at the very smallest scale to create a very different type of computer. Where a conventional computer stores information in "bits" in two states, either 1 or 0, a quantum computer uses "qubits" to operate in multiple states at the same time. This ability to encode information in parallel enables quantum computers to solve complex problems in hours or days that would take billions of years using today's computing techniques.

At Microsoft, we're taking a three-pronged approach to quantum computing research and development. First, we are pioneering a revolutionary "topological" approach to quantum computing that performs computations longer and more consistently with significantly lower error rates compared with other quantum methods. We've also made significant investments in building a global team of researchers who are working at the forefront of quantum research. The team includes some of the leading mathematicians, physicists and computer scientists in the world, bolstered by collaborations with many other across universities and industry. Finally, we are working on every aspect of the quantum computing challenge with the goal of developing a scalable end-to-end quantum computer. This includes developing hardware, software and developer programming tools needed to make quantum computing a reality.

For further information, please visit:
https://www.microsoft.com/en-us/quantum/

Chapter 4

Microsoft's commitment to a trusted, responsible and inclusive cloud

Microsoft's commitment to a trusted, reponsible and inclusive cloud

For Microsoft, creating a cloud for global good starts with a commitment to harnessing our own resources and being accountable for what we do. We believe a trusted, responsible and inclusive cloud is grounded in how we engage as a business, the development of our technology, our advocacy and outreach, corporate philanthropy, and how we are serving the communities in which we operate.

At Microsoft, our commitments include:

Increased transparency. We believe that providing clear and relevant information about the issues covered in this document to customers, business partners, governments and others is vital to the work of creating a cloud for global good. We will continue to focus on expanding transparency; one example of this is our Transparency Hub. On this site, we disclose key information about our operations, including environmental and workforce data, details about our supply chain and political engagement, and summaries of requests we receive for customer data from governments.

Focused advocacy. We are dedicated to using our voice and resources to push for the changes needed to deliver the benefits of the cloud to people around the world. Microsoft operates in more than 120 countries, and we have deep economic and social connections to the communities in which we live and work. We aspire to use our knowledge of local conditions in combination with our global experience to drive informed and sustainable policy decisions that serve the interests of our customers, local communities and, ultimately, the global good.

Strong partnerships. We will continue to work with governments, civil society and industry on projects and programs designed to ensure that the benefits of cloud computing are available to all. We will also focus on increasing inclusion and expanding empowerment to people who still lack access to technology and the opportunities it enables. Building on the foundation of Microsoft's long history of corporate giving, we will seek new ways to improve outcomes for more and more people around the world.

Constructive conversations. We believe that the best — and only — way to realize the opportunities that cloud computing offers is through ongoing and inclusive discussions. We'll use our resources to bring interested parties together to talk about how to address the challenges that we all face. And we will continue to create platforms where people and organizations at the local, regional and global levels can raise concerns, share their best ideas and work toward solutions.

Ongoing research. We'll work closely with leading researchers and academics to develop and share additional insights on cloud computing. To help policymakers understand the complex legal and economic implications of existing and emerging technology innovations, we will continue to facilitate evidence-based decision-making processes and support comprehensive and independent research across a broad range of disciplines and policy issues.

In the following section, we will outline how Microsoft is living up to these commitments by highlighting a few of the investments we have made in the past year to create a more trusted, responsible and inclusive cloud.

oud for Global Goo

Building a trusted cloud

A Digital Geneva Convention

Protecting cyberspace in times of peace

Across the world, governments continue to invest in greater offensive capabilities in cyberspace, and nation-state attacks on civilians and critical infrastructure are on the rise. Some 74 percent of the world's businesses are expecting to be hacked each year, with the economic cost of cybercrime estimated to reach U.S. $3 trillion by 2020.

Yet, the financial cost is just one element of the cyberthreat challenge. Many of the attacks expose personal data, or spread misinformation online. Moreover, online threats do not originate with criminals alone. In the past few years we have witnessed an increase in government investment in offensive cybercapabilities, and as a result a greater number of cyberattacks that appear to be backed by nation-states. These range from the Wannacry attack that held to ransom over 100,000 computers in hospitals, manufacturing plants, education systems and logistics operations across the world to the NotPetya attack, which appeared to target critical infrastructure in the Ukraine but affected companies around the world.

At Microsoft we accept that no single measure will be sufficient to address the cybersecurity challenges we face, and indeed that they will never go away completely. However, the events of the past year underscore that the time has come for the world's governments to agree to a set of rules of behavior to ensure long-term stability of the online environment. In particular, governments must refrain from attacking civilians and critical infrastructures in cyberspace.

For this reason, in early 2017, Microsoft called for the creation of a Digital Geneva Convention. This convention would commit governments to adopt and implement norms that have been developed to protect civilians on the internet, without introducing restrictions on online content.

A Digital Geneva Convention:

The Fourth Geneva Convention has long protected civilians in times of war. It is now clear that we need to build on this legacy and forge a Digital Geneva Convention that will commit governments to protecting civilians from nation-state attacks in times of peace.

The foundations for this effort are already in place, but it took the international community a long time to reach the barest of consensus — that international law applies in cyberspace. The United Nations almost two decades ago set up a working body to ensure agreement is reached on how to handle the then relatively new field of information technology (IT), and in particular the increasingly difficult question of cybersecurity. It was only in 2015 that the United Nations Group of Governmental Experts on Developments in the Field of Information and Telecommunications in the Context of International Security (UNGGE) confirmed that international law applies to cyberspace. Although particular groupings, such as G7 and G20, have reasserted this position, it seems progress has stalled.

All of this points to the need to identify new steps ahead. These could include:

Governments should pursue a multilateral agreement that affirms recent cybersecurity norms as global rules. Just as the world's governments came together in 1949 to adopt the Fourth Geneva Convention to protect civilians in times of war, we need a Digital Geneva Convention that will commit governments to implement the norms that have been developed to protect civilians on the internet in times of peace. More than that, the drafting process should be used to add specificity to the broad agreements in place now and ensure their implementation. For example, such a convention should commit governments to avoiding cyberattacks that target the private sector or critical infrastructure, or the use of hacking to steal intellectual property. Similarly, it should require that governments assist private sector efforts to detect, contain, respond to and recover from these events, and should mandate that governments report vulnerabilities to vendors rather than stockpile, sell or exploit them.

There is a need to set up an independent attribution organization that spans the public and private sectors. Today a perception exists about the ability to accurately attribute cyberattacks to their perpetrators. It is time for industry to dispel this myth. Although not simple, attribution capabilities have improved dramatically over the past few years. However, the capabilities are often dispersed between different technology companies and governments, and there are no established rules to dictate how this information could be shared. To address this, the world needs an independent organization that can investigate and share the evidence that attributes nation-state attacks to specific countries. Although there is no perfect analogy, the world needs an organization that can address cyberthreats in a manner like the role played by the International Atomic Energy Agency in the field of nuclear nonproliferation. This organization should consist of technical experts from across governments, the private sector, academia and civil society with the capability to examine specific attacks and share the evidence showing that a given attack was by a specific nation-state. Only then will nation-states know that if they violate the rules, the world will learn about it.

The tech sector needs to act collectively to better protect the internet and customers everywhere from nation-state attacks. As the first responders to threats that in part target our own infrastructure, it's important for global technology companies to adopt concrete commitments to help deter and respond to nation-state cyberattacks. We believe that should include a pledge not to assist any actor, including governments, in attacking the information infrastructure of another party, irrespective of where they are in the world; a pledge to work together to address security issues, as well as to not traffic in vulnerabilities for offensive purposes or embrace business models that do so.

What Microsoft is doing:

The tech sector plays a unique role as the internet's first responders, and we should commit ourselves to collective action that will make the internet a safer place, affirming a role as a neutral Digital Switzerland that assists customers everywhere and retains the world's trust. Microsoft is aggressively taking new steps to better protect and defend customers. This includes new security features at every level of the technology stack, reflecting the $1 billion that we're spending annually in the security field.

Within Microsoft we've forged a unique, internal three-part partnership among the 3,500 security professionals from across the company. The Microsoft Threat Intelligence Center (MSTIC) is our reconnaissance arm, combing through the constant stream of data from our more than 200 cloud services and third-party feeds. Using machine learning, behavioral analysis and forensic techniques, this dedicated team creates a real-time picture — a security intelligence graph — of cyberactivity related to advanced and persistent threats to Microsoft and our customers. When a threat is detected, MSTIC alerts our Cyber Defense Operations Center (CDOC), an "eyes on glass" command center staffed 24 hours a day, seven days a week by rotating teams of security and engineering professionals from across our product and services portfolio. This team of specialists serves as our front line, taking immediate action against threats to defend our own systems and protect customers.

As we identify threats, we're not only working with customers, but using legal process, led by our Digital Crimes Unit (DCU), to respond in new and innovative ways that disrupt attacks, including those launched by nation-states. Last year MSTIC identified an attack pattern that led to a group associated with a nation-state that had registered internet domains using names that included Microsoft and other companies' trademarks. We went to federal court, obtained court orders and successfully sought appointment of a special master to oversee and expedite additional motions in our case. Working under this judicial supervision, we can notify internet registries whenever this group registers a fake Microsoft domain and request that control of that domain be transferred immediately to a sinkhole operated by DCU.

Using this novel approach, we can disrupt the nation-state's use of these domains within 24 hours. Since last summer, in response to extended nation-state attacks, we have taken down 60 domains in 49 countries spread over six continents. In each instance we stopped the flow of data to the hackers from any customers whose computers were hacked, we notified the customers of the nation-state attack, and we helped them clean their environment and increase their security.

We are also working hard to protect other areas of potential vulnerability in the digital estate. Email, for example, is responsible for an estimated 90 percent of all hacking via "phishing" attacks. In response to this, we have introduced Advanced Threat

Protection for Microsoft Exchange Online. It identifies recognizable malware and suspicious code patterns in emails and stops them before they can do damage. In addition, Office 365 Threat Intelligence provides enterprises with information on the top targeted users, malware frequency and security recommendations related to their business. Building on that, we added new data governance features for Office 365, including alerts that will be sent automatically to users when someone attempts to copy and download their inbox. We'll be adding new features and offers in the coming months that provide additional protection.

But security-related product features are just the start. Data analytics and machine learning have become game-changing defense mechanisms. Microsoft's datacenters are connected to over a billion computing end points and receive over a trillion data points every day. Advanced Threat Protection alone processes 6 billion emails each day. This provides the foundation for world-class early warning systems to detect cybersecurity attacks.

Across the tech sector, companies are racing to provide stronger cybersecurity protection for customers, including from nation-states. Each of our advances is making an important contribution. But we're nowhere close to being able to declare victory. We therefore need to recognize a critical truth — this is not a problem that we can solve solely with each of us acting alone.

If we're going to turn these words into effective action, we need to come together as an industry to develop our own clear principles and to help put in place the steps needed to make these principles real. For example, we should commit ourselves to collaborative and proactive defense against nation-state attacks and to remediate the impact of such attacks. We should pledge that we'll continue to take no efforts to assist in offensive actions anywhere. We should make software patches available to all our users, regardless of the attackers and their motives. We should adopt coordinated disclosure practices for the handling of product and service vulnerabilities.

We need to come together as an industry to develop our own clear principles.

There is strong progress on which we can build. For example, we at Microsoft have been collaborating with other leading cloud companies like Amazon and Google to combat cloud abuse such as spam and phishing sites. We're working together on a common abuse reporting schema to accelerate the reporting of abuses we may see on each other's networks. On issues such as customer notification of potential nation-state attacks, we've all learned from important work where Google and Facebook have been early and impressive leaders. More broadly, there is good work and common collaboration springing up everywhere, from new startups to the industry's largest companies.

A Digital Geneva Convention will safeguard citizens around the world from major state-led or state-sanctioned cyberattacks. Given the critical importance of effective cybersecurity to international peace and the stability of the global economy, such an initiative is becoming increasingly critical. We need to build on the work done to date, but move with greater urgency and take steps now to pave the way a legally binding agreement that will ensure a stable and secure cyberspace. Microsoft is committed to working with governments around the world and the global tech industry to find a practical way forward to make this vision a reality.

Building a responsible cloud

AI for Earth

We face major challenges as a society in the way in which we manage the impact of our activity on our environment. If we are to enjoy the benefits of a healthy, functioning planet, we must find a way to use the world's resources in a way that allows us to live and work in a more sustainable fashion.

This is no small challenge. With the world's population set to reach 10 billion by 2050, we will need to produce 70 percent more food on less land, using less water and without resorting to a significant increase in the use of existing fertilizer. Global biodiversity continues to decline with species extinction rates at record levels, and projected demand for fresh water over the next five decades is projected to outstrip supply by 40 percent. Climate change is exacerbating the speed, magnitude and severity of these issues, and causing dramatic global changes in our ecosystems that threaten human health, infrastructure and natural systems.

The scale and speed of the changes we see in our physical and natural world require new solutions. But the latest innovative technologies often come with a price tag and require computational expertise that puts them out of reach for many researchers and nongovernmental organizations. In response, Microsoft has created AI for Earth.[50] The program is aimed at putting the power of AI toward solving some of the biggest environmental challenges of our time. In late 2017, we announced an expansion of the program, including an additional $50 million in funding.

The program has three pillars:

Access: We will improve access by making a new pool of grants available to help researchers and organizations gain access to cloud and AI computing resources. This includes access to Azure compute time and our data science virtual machine offerings on Azure. These grant applications are available today.

Education: We will provide new training and educational opportunities to make sure people and organizations know what AI tools are available, how to use them and how the tools can help meet their specific needs. Our approach will be both broad and deep, reaching many people through general session trainings as well as small-group faculty summits on single-issue areas and training for grantees.

Innovation: We also want to encourage others to innovate based on the power and potential of AI. We will partner with others on lighthouse projects that demonstrate how AI can deliver results more rapidly, accurately and efficiently. Already, we have three projects underway — one enabling land cover mapping to aid precision conservation; another that will enable smart agriculture through sensors, drones, data and broadband connectivity; and another that will test the viability of using our smart mosquito traps to remotely track and monitor species health.

AI for Earth builds upon Microsoft's long history of innovation in AI, as well as our commitment to sustainability. We've taken many steps to operate more sustainably as a company, including operating 100 percent carbon neutral since 2012, setting commitments to increase the amount of renewable energy we use to power our datacenters and ensuring they operate more efficiently, and using the cloud and AI within our operations to reduce our resource consumption across the globe.

The private sector has an important role to play in addressing climate change, both within our operations and by democratizing the use of advanced technologies like AI across companies, countries and research organizations of all sizes, on every continent. These tools, and the insights derived from their use, can help our customers and partners not only reduce emissions to slow or stop climate change, but also adapt and thrive in a changing environment.

Building an inclusive cloud

Closing the rural broadband gap in America

America is a vast country that continues to inspire inventors and innovators to devise and advance new ways to bring its people closer together and drive its economy forward. This can be seen in wave after wave of new technologies that have helped revolutionize transportation, communication and commerce for this country, and for the world. But the speed and spread of these innovations has often been uneven, sometimes leaving entire groups of Americans behind.

Access to electricity — the foundation of so many of the key innovations of modern life — is one example. By the mid-1930s, the electrification of urban America was nearly complete and had already revolutionized manufacturing and transformed almost every aspect of people's day-to-day lives at work and at home.

But for millions of Americans living in rural towns, life remained largely unchanged. In 1930, only 10 percent of the country's 6 million farms were connected to the electric grid. That meant the average farming family still spent 10 hours a week hauling water. Cows were milked by hand. Modern conveniences that city residents took for granted — refrigeration, electric lights, indoor plumbing — remained a dream for their rural fellow citizens.

Over the next 20 years, a nationwide focus on rural electrification brought this innovation to millions who had previously been left behind. By 1950, 90 percent of American farms were wired for electricity. The impact on economic opportunity and quality of life was dramatic: Automated milking machines cut the labor required to collect milk by 50 percent; the average value of crops per farm jumped more than 34 percent; and farming families no longer had to heat water for cooking over wood fires, light their homes with kerosene lamps or wash clothes by hand.

We've reached a similar moment in 21st century America. For so many of us, computers, mobile devices and cloud computing have already transformed how we connect to one another, work, learn and play. And we are only at the beginning of a sweeping technology revolution that offers the promise of new economic opportunities and new ways to address a wide range of once-unsolvable problems.

But as was true in the 1930s, rural Americans face significant barriers when it comes to accessing the benefits of the current technology revolution. Although 90 percent of Americans have broadband access, nearly 40 percent of Americans living in rural areas don't live within reach of a broadband connection — 23.4 million rural Americans in all — meaning they are unable to take advantage of the economic and educational opportunities enjoyed by their urban neighbors.[51][52]

Yet despite this glaring disparity, real progress to close the rural broadband gap has plateaued in recent years. High costs, the absence of new and alternative technologies, and market and regulatory conditions have hampered efforts to expand coverage. But this is changing, thanks to recent advancements in technology, newly adopted standards, business model innovations and a growing demand for a broad range of cloud services.

A new rural broadband strategy

In July 2017, Microsoft called for the elimination of the rural broadband gap in America within the next five years.

We believe that this is an achievable goal based on a new strategic approach that combines private-sector capital investments focused on new technologies with public-sector support. This is supported by findings by the Boston Consulting Group suggesting that a combination of technologies can substantially reduce the total cost of extending broadband coverage.

Specifically, a technology model that uses a combination of the TV White Spaces spectrum, fixed wireless and satellite coverage can reduce the initial capital and operating costs by roughly 80 percent compared with the cost of using fiber cables alone, and by approximately 50 percent compared with the cost of current LTE fixed wireless technology.

One key to deploying this strategy successfully is to use the right technology in the right places.

TV White Spaces is expected to provide the best approach to reach approximately 80 percent of this underserved rural population, particularly in areas with a population density between two and 200 people per square mile. Microsoft itself has considerable experience with this technology, having deployed 20 TV White Spaces projects worldwide.

But TV White Spaces alone will not provide the complete solution. Satellite coverage is expected to be the most cost-effective solution for most areas with a population density of less than two people per square mile, and LTE fixed wireless for most areas with a density greater than 200 people per square mile. This mixed model for expanding broadband coverage will likely bring the total national cost of closing the rural broadband gap to roughly $10 billion.

Microsoft's new Rural Airband Initiative

At Microsoft, we're prepared to invest our own resources to help serve as a catalyst for broader market adoption of this new model. We're committed to three elements on a five-year basis:

1. Direct projects with partners.

Microsoft will invest in partnerships with telecommunications companies with the goal of bringing broadband connectivity to 2 million people in rural America by July 4, 2022. We and our partners will have 12 projects up and running in 12 states in the next 12 months.

Our goal is not to enter the telecommunications business ourselves or to profit directly from these projects. We will invest in the upfront capital projects needed to expand broadband coverage, seek a revenue share from operators to recoup our investment, and then use these revenue proceeds to invest in additional projects to expand coverage further.

2. Digital skills training for people of all ages.

Working through Microsoft Philanthropies, our Rural Airband Initiative will invest in helping train people of all ages in these rural communities on the latest technologies so they can use this new connectivity to improve education, healthcare and agriculture, and transform their businesses.

Our first partnership under the Rural Airband Initiative will be a multiyear partnership with National 4-H Council — engaging America's largest youth development organization, 4-H, to provide digital literacy skills training to youth as well as teen-led learning programs in rural communities.

3. Stimulating investment by others through technology licensing.

Our ultimate goal is to help serve as a catalyst for market investments by others in order to reach additional rural communities. That's why we're launching a new program to stimulate investment through royalty-free access to at least 38 patents and sample source code related to technology we've developed to better enable broadband connectivity through the use of TV White Spaces spectrum in rural areas.

A vital role for the public sector

Although we believe the private sector can play the leading role in closing the rural broadband gap, the public sector also has a vital role to play. Three related governmental measures are needed:

First, the Federal Communications Commission (FCC) needs to ensure the continued use of the spectrum needed for this mixed technology model. Specifically, it will be important for the FCC to ensure that at least three channels below 700 MHz — the so-called TV White Spaces — are available for wireless use on an unlicensed basis in every market in the country, with additional TV White Spaces available in smaller markets and rural areas.

In addition, federal and state infrastructure investments should include targeted funds on a matching basis for the capital investments that will best expand coverage into rural areas that currently lack broadband access. These funds should be made available for use by multiple technologies based on what is most needed.

What's at stake

It's not just the United States that has the challenge of a broadband gap. Across the world, there are billions of people who have no access to high-speed broadband or who struggle with the affordability of connectivity. But, like many countries, America has become accustomed to ongoing capital investments to expand broadband capacity in areas that already have broadband coverage. The time has come to expand this coverage to those areas that currently lack it entirely.

And although we are making a major push in the United States, the Microsoft's Rural Airband Initiative is drawing on years of experience from more than 20 projects in 10 countries. This experience has given us the insights to know where to put our resources as well as where we need the support and expertise of others.

We believe there is an opportunity for other companies large and small to join in with market-based investments. We all have the opportunity to innovate together — achieving together what none of us can accomplish alone.

And just as we look forward to sharing what we have learned as a company, we look forward to applying over the next five years what we undoubtedly can learn from others.

Broadband connections have become indispensable for accessing healthcare, advancing education, improving agriculture and growing a small business. No country should settle for an outcome that leaves behind a large percentage of the community. We can and should bring the benefits of broadband coverage to every corner of the nation.

We look forward to working in partnership with government leaders at all levels, private-sector companies that have the expertise to develop and deliver affordable solutions, and local community members who can help enable the capabilities that a new generation of digital innovations and cloud computing can provide.

Conclusion

Working together to create a cloud for global good

Significant advances in public policy have accompanied every industrial revolution. From labor laws to education policy, from security standards to privacy regulations, and from universal access to competition policy, each industrial age has seen the emergence of new policy frameworks that balance the opportunities and challenges that are ushered in with any technological leap. This cloud-enabled revolution is no different.

But policy is not the role of governments alone. The development and implementation of new policies requires collective input and action from across the public and private sectors, working in close partnership with civil society and all those who are the most affected by these developments.

If we are to truly build a cloud for global good, it will be essential for governments, citizens, businesses and organizations to work together to create policy frameworks that ensure that no one is left behind and no one is put at risk. We need policies that will guide the development of solutions to solve some of world's most pressing problems and unleash innovation from individuals and communities across the planet.

At the same time, we must accept that — as was the case with every previous industrial revolution — this digital transformation will come with disruption and dislocation. But with this revolution, we have the opportunity to draw on the lessons of history to guide our response. If we learn from the examples of the past, we can design the policy frameworks needed to help us deliver a new generation of technology innovation that will benefit all.

The digital transformation stories in this book inspire us, illustrating the way in which people are using technology to improve people's lives and lift communities around the world.

They remind us of the importance of continuing the work of shaping the laws, regulation, policies and practices that will help unlock the potential of cloud-enabled technologies while protecting people against the inevitable challenges that will arise.

If we are to truly build a cloud for global good, it will be essential for governments, citizens, businesses and organizations to work together.

As we have said before, this moment in history demands a framework of laws and regulations created by people represented through their governments that preserves and protects important values and provides fair and equal access to benefits that only a trusted, responsible and inclusive cloud can deliver.

More than ever, we need to work together to achieve this, and we look forward to continuing to engage with policymakers, business owners, educators, citizens, advocates, experts — everyone who has a stake in the outcome — to craft a framework for cloud computing that will help us create a cloud for global good.

G 08 01

Spotlight

A look into mixed reality

Mixed reality is the result of blending the physical world with the digital world. It is the next evolution in human, computer and environment interaction, and unlocks possibilities that before now have been restricted to our imaginations. It is made possible by advances in computer vision, graphical processing power, display technology and input systems. The term "mixed reality" was originally introduced in a 1994 paper by Paul Milgram and Fumio Kishino, "A Taxonomy of Mixed Reality Visual Displays." Since then, the application of mixed reality has expanded beyond displays to include environmental input, spatial sound and location.

Over the past several decades, the relationship between human input and computer input has been well explored. It even has a widely studied discipline known as "human-computer interaction" or HCI. Human input happens through a variety of means including keyboards, mice, touch, ink, voice and even Kinect skeletal tracking.

Advances in sensors and processing are giving rise to computer input from environments. **The interaction between computers and environments is creating a new form of interaction based on computer perception.** This is why the API names in Microsoft Windows that track environmental information are called perception APIs. Environmental input captures things like a person's position in the world (e.g., head tracking), surfaces and boundaries (e.g., spatial mapping and spatial understanding), lighting, environmental sound, object recognition, and location.

Now, the combination of computer processing, human input and environmental input is opening the door to new opportunities to create true mixed-reality experiences. Movement through the physical world can translate to movement in the digital world. Boundaries in the physical world can influence application experiences such as gameplay. Without environmental input, experiences cannot blend physical and digital realities.

Since mixed reality is the blending of the physical world and digital world, these two realities define the polar ends of a spectrum known as the "virtuality continuum." For simplicity, we refer to this as the "mixed-reality spectrum." On one end of the spectrum is physical reality in which humans exist. On the other is the corresponding digital reality.

Most mobile phones today have little or no environmental understanding capabilities. As a result, they can't offer mixed-reality experiences. Experiences that overlay graphics on video streams of the physical world are augmented reality. Experiences that occlude your view to present a digital experience are virtual reality. Experiences between these two extremes are mixed reality, including, for example, the projection of holograms onto the physical world, or the representation of real-world objects, like the walls of your living room, in the digital world.

Microsoft is working on a number of mixed-reality initiatives, most notably the Microsoft HoloLens, the first self-contained holographic computer, enabling you to interact with holograms in the world around you.

A demonstration of the HoloLens is available here: https://www.microsoft.com/en-us/hololens.

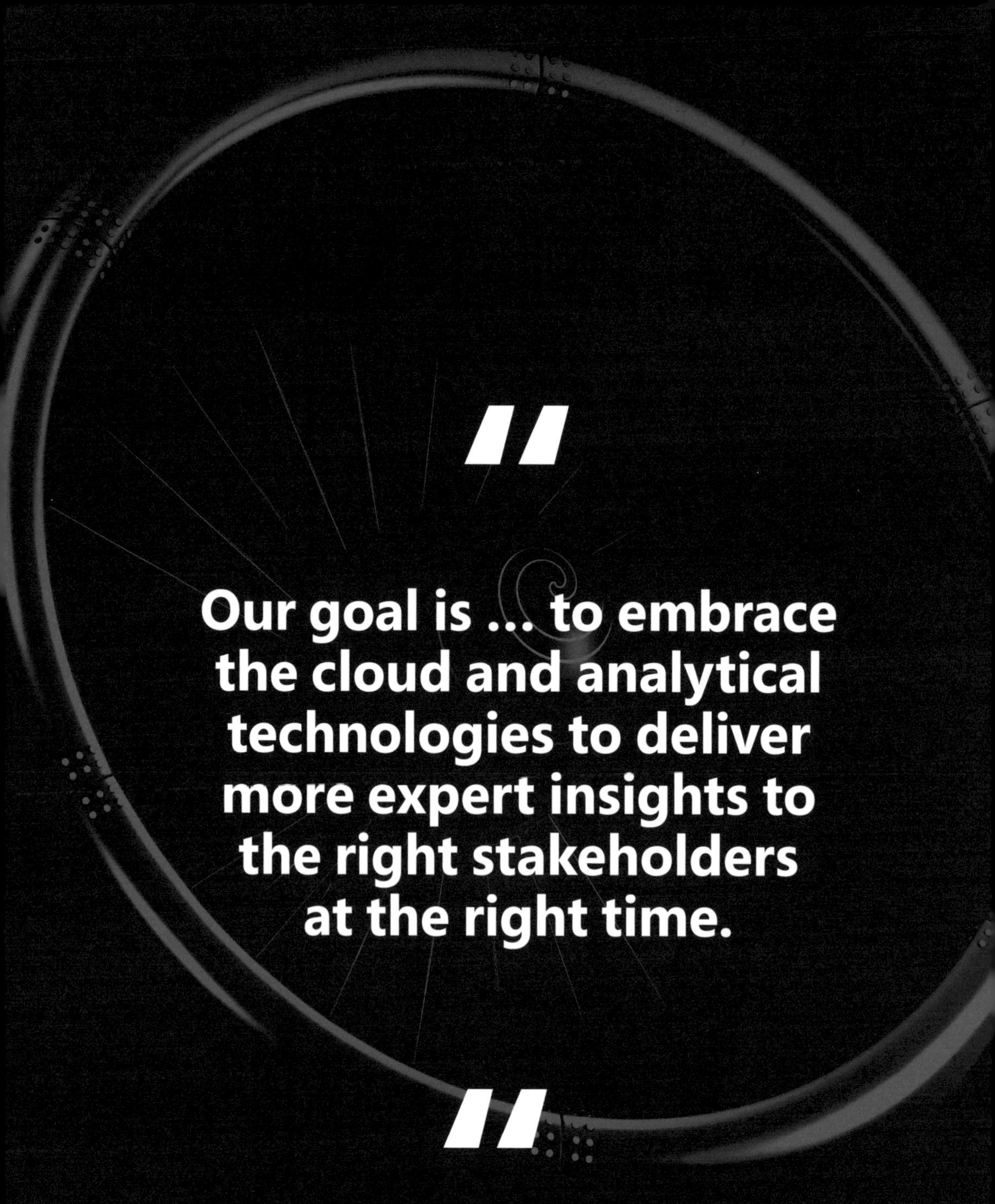

"

Our goal is ... to embrace the cloud and analytical technologies to deliver more expert insights to the right stakeholders at the right time.

"

Rolls-Royce:

Fueling engine efficency with the cloud

Challenge

Worldwide, flight delays and disruptions cost the airline industry billions of dollars every year. Just one unscheduled delay and its ripple effect on the fleet and passengers can add up to a million dollars a day. Similarly, fuel costs generally account for a whopping 40 percent of airlines' operating expenses. Even a 1 percent optimization of fuel consumption can save a carrier millions of dollars annually.

To reduce costs and improve their margins, airlines constantly search for ways to improve operational efficiency. With modern airplane engines equipped with thousands of sensors generating terabytes of data, advanced cloud technology offers new opportunities to do just that.

Technology solution

Airplane engine manufacturer Rolls-Royce offers comprehensive maintenance services to carriers that use its engines. To improve customer service and streamline operations, the company employs advanced analytics, combining sensor data and other information to maximize aircraft availability and fuel efficiency for more than 13,000 commercial aircraft engines in service.

Fueling engine efficency with the cloud

Using the Microsoft Azure Platform, Azure IoT Suite and the Cortana Intelligence Suite, Rolls-Royce is collecting, aggregating and analyzing data from disparate sources at unprecedented scale. In doing so, the company is helping its customers pinpoint operational anomalies and optimize aircraft performance.

Outcome

With the ability to analyze vast amounts of data, including engine performance snapshots, technical logs, flight plans and weather information, Rolls-Royce is improving the reliability and efficiency of air travel. By comparing information from specific aircraft components with data models and other components in the fleet, for example, Rolls-Royce can alert flight crews when a piece of equipment is underperforming or due for replacement.

Rolls-Royce also can analyze fuel data to help airlines understand exactly which factors — including flight plans, equipment maintenance, weather and discretionary fuel — have the most impact on fuel performance.

"Our goal is ... to embrace the cloud and analytical technologies to deliver more expert insights to the right stakeholders at the right time," says Nick Farrant, senior vice president at Rolls-Royce. "If we can do that and link new digital capabilities into our services, we can collaborate more deeply with our customers and solve many more of their problems."

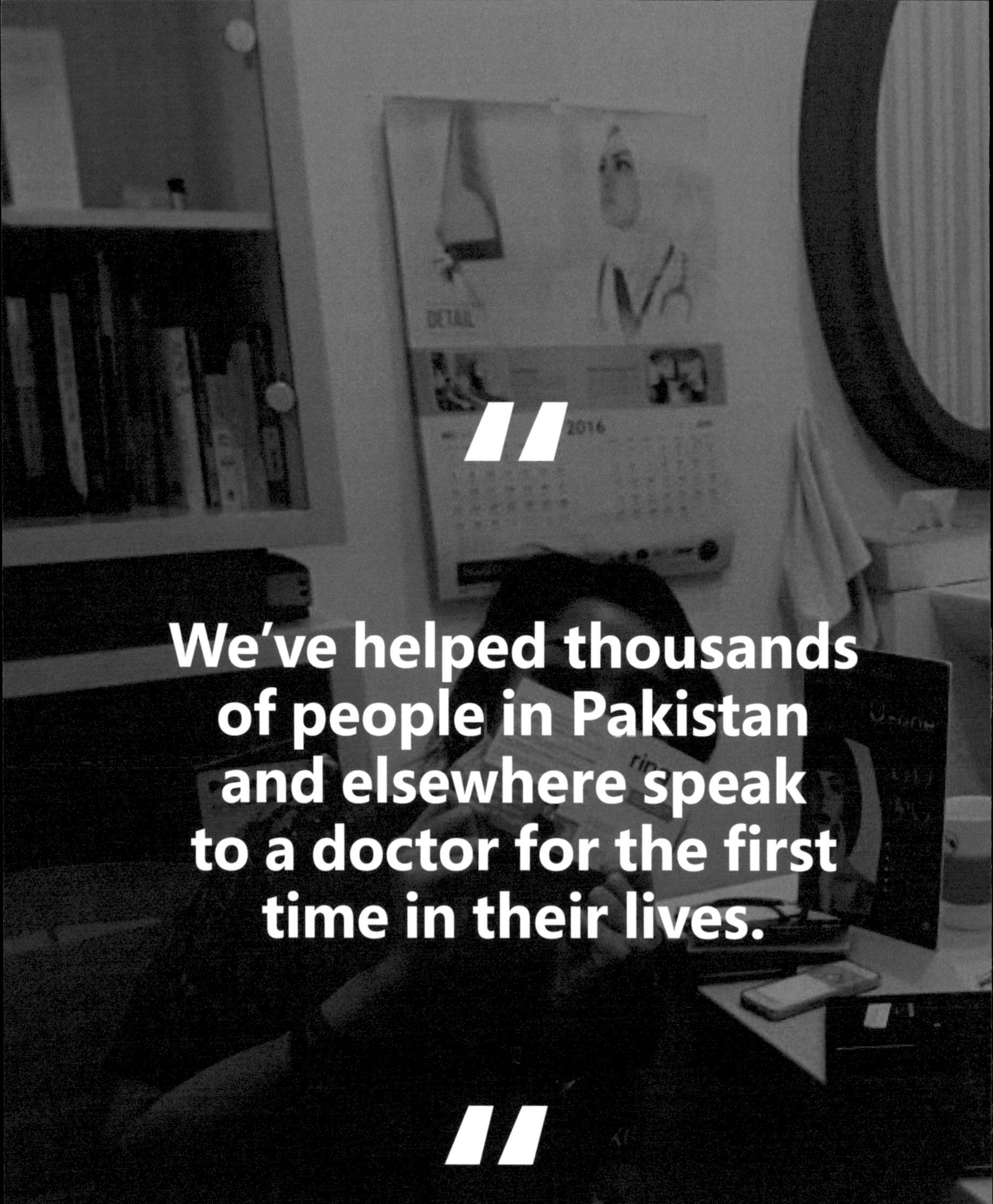

“We’ve helped thousands of people in Pakistan and elsewhere speak to a doctor for the first time in their lives.”

RingMD:

Connecting patients with doctors, virtually anywhere

Challenge

Countless people around the world lack access to medical care, due to proximity, economics, cultural norms or other barriers. India, for example, has fewer than one doctor per 1,000 individuals, according to the World Health Organization. For people in rural areas, traveling to and from the nearest hospital by bus can take half a day or longer. Many don't have the time or money, while others are too ill for the journey.

Technology solution

During a trip to Indonesia, Justin Fulcher was taken aback when he saw the dichotomy of a man drinking water scooped from the ground while holding a mobile phone. Inspired, Fulcher founded RingMD, using cloud technology to help fill the gaps he perceived in the global healthcare ecosystem while increasing access and affordability for people in need. RingMD connects doctors and patients through an app that facilitates online video consultations, allowing users to speak with medical professionals and exchange medical files — anytime and from anywhere.

Connecting patients with doctors, virtually anywhere

The platform is hosted by the Microsoft cloud, allowing it to scale much more quickly and cost-effectively than traditional health systems, which are reliant on physical infrastructure and on-site personnel. RingMD also employs machine learning, AI and complex algorithms that enable it to provide patients with customized health recommendations.

Outcome

RingMD now operates in nine languages across 10 countries, including India, Pakistan and Thailand. It brings together healthcare professionals and patients across great distances at a fraction of the cost of conventional health services. And it has been a boon for practitioners like Dr. Vidya Neelkant, as well as the patients she treats virtually. She's encouraged by the medical advice she's able to provide using RingMD, particularly to women.

"Women in India don't like to see male doctors, and they often live so far away from a hospital that they don't get the care they need, especially when they're pregnant," Neelkant said. "It's so good that I can help them have healthier pregnancies now and advise them on things like diet, hygiene and how much to work. A mother's health is very important for her baby, yet not everyone is getting this care, especially in the rural areas."

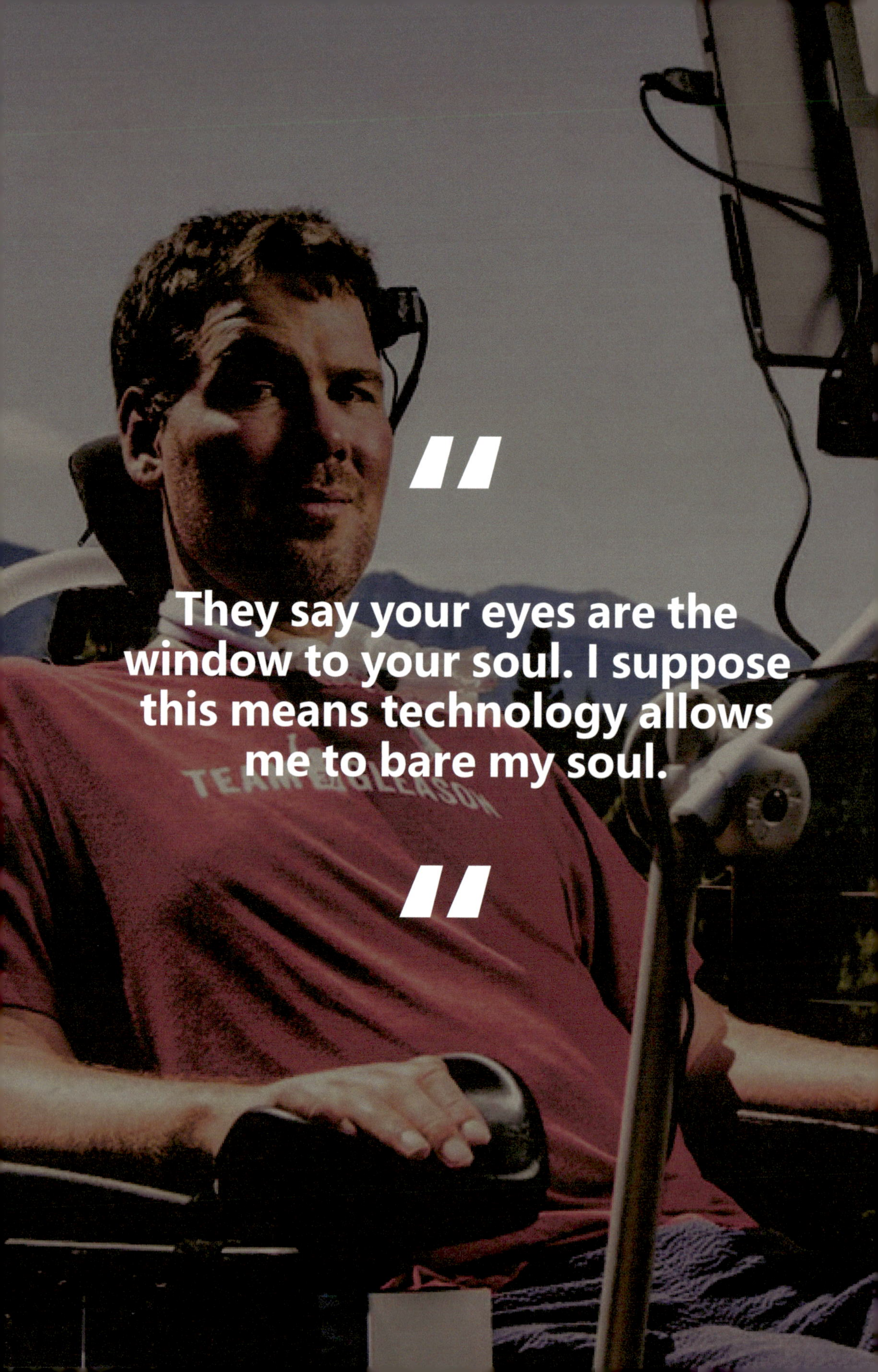
"
They say your eyes are the window to your soul. I suppose this means technology allows me to bare my soul.
"

Team Gleason:

Using technology to empower people with ALS

Challenge

In the NFL, Steve Gleason was inspiring and relentless. The undersized player made it as a safety with the New Orleans Saints. And he became a local icon after making a remarkable play that helped unite the city following the devastation of Hurricane Katrina.

Today, facing even longer odds, Gleason continues to unite and galvanize others. He was diagnosed in 2011 with ALS, or Lou Gehrig's disease, and has been stripped of his ability to speak, move or even breathe unassisted. The disease ravages nerves in the brain and spine, and there is no known cure. But despite his diagnosis, Gleason is determined to lead a full life, be an active husband and father, and help others with ALS.

Technology solution

Following his diagnosis and the onset of symptoms, Gleason formed Team Gleason and focused on combating ALS on a greater scale. He engaged the technology industry to help make a tangible difference in the lives of those with the disease. Recognizing an opportunity to empower people through AI, innovators at Microsoft were eager to participate.

Using technology to empower people with ALS

On the eve of Microsoft's first annual hackathon, //oneweek, Gleason issued a challenge: Develop technology that helps people with ALS communicate and move more easily. That challenge inspired Eye Control for Windows 10, enabling people with limited mobility and speech to move their wheelchairs and communicate through an AI-assisted user interface and keyboard operated by eye movement.

Outcome

In addition to helping and inspiring others diagnosed with ALS, Gleason had personal goals when he reached out to the technology industry. He wanted the ability to play with his young son, talk more easily with his wife, move his wheelchair on his own, and independently power his Microsoft Surface on and off.

Gleason's //oneweek team achieved all four objectives with their hack for his wheelchair; they won the Hackathon, too. Now, using Eye Control and a Surface Pro 3 tablet, Gleason can input sentences, communicate with friends and family, and interact with his son from his wheelchair — just as he envisioned. Moreover, eye-tracking support has been built into Windows 10 to help countless others use their machines to stay active and connected.

"Until there is a medical cure for ALS," Gleason says, "technology will be that cure."

Despite ALS diagnosis, NFL icon sees bright future with Eye Control

Endnotes

1. Manyika, James; Woetzel, Jonathan; Dobbs, Richard; Remes, Jaana; Labaye, Eric; and Jordan Andrew. "Can long-term global growth be saved?" McKinsey Global Institute Report, McKinsey & Company, January 2015, https://www.mckinsey.com/global-themes/employment-and-growth/can-long-term-global-growth-be-saved.

2. "Cancer fact sheet." World Health Organization, February 2017, http://www.who.int/mediacentre/factsheets/fs297/en.

3. Purdy, Mark, and Daugherty, Paul. "Why Artificial Intelligence is the Future of Growth." Accenture, 2016, https://www.accenture.com/us-en/_acnmedia/PDF-33/Accenture-Why-AI-is-the-Future-of-Growth.pdf.

4. Richards, Kevin; LaSalle, Ryan; Devost, Matt; van den Dool, Floris; and Kennedy-White, Josh. "2017 Cost of Cyber Crime Study: Insights on the Security investments That Make a Difference." Accenture and Ponemon Institute, 2017, https://www.accenture.com/t20170926T072837Z__w__/us-en/_acnmedia/PDF-61/Accenture-2017-CostCyberCrimeStudy.pdf.

5. Purdy, Mark, and Daugherty, Paul. "Why Artificial Intelligence is the Future of Growth." Accenture, 2016, https://www.accenture.com/us-en/_acnmedia/PDF-33/Accenture-Why-AI-is-the-Future-of-Growth.pdf.

6. Statt, Nick. "iPhone manufacturer Foxconn plans to replace almost every human worker with robots." The Verge, December 30, 2016, https://www.theverge.com/2016/12/30/14128870/foxconn-robots-automation-apple-iphone-china-manufacturing.

7. Stubbings, Carol; Williams, Jon; Sethi, Bhushan; and Brown, Justine. "Workforce of the future: The competing forces shaping 2030." PwC, 2017, https://www.pwc.com/gx/en/services/people-organisation/publications/workforce-of-the-future.html.

8. Smith, Brad. "Moving forward together: Our thoughts on the US election." Microsoft On the Issues, Microsoft, November 9, 2016, https://blogs.microsoft.com/on-the-issues/2016/11/09/moving-forward-together-thoughts-us-election.

9. Tetlow, Gemma. "Blame technology not globalisation for rising inequality, says IMF." Financial Times, April 10, 2017, https://www.ft.com/content/cfbd0af6-1e0b-11e7-b7d3-163f5a7f229c.

10. Stewart, Michael. "Trust in Tech: No Room for Complacency." Research Insight, Edelman, March 16, 2017, https://www.edelman.com/post/trust-tech-no-room-complacency.

11. Smith, Brad. "The need for a Digital Geneva Convention." Microsoft On the Issues, Microsoft, February 14, 2017, https://blogs.microsoft.com/on-the-issues/2017/02/14/need-digital-geneva-convention.

12. AI for Earth, Microsoft, 2017, https://www.microsoft.com/en-us/aiforearth.

13. Smith, Brad. "A rural broadband strategy: connecting rural America to new opportunities." Microsoft On the Issues, Microsoft, July 10, 2017, https://blogs.microsoft.com/on-the-issues/2017/07/10/rural-broadband-strategy-connecting-rural-america-new-opportunities.

14. "RGS Statement on US-UK Data Protection Discussions." Reform Government Surveillance, tumblr, July 15, 2016, https://reformgs.tumblr.com/post/147464333157/rgs-statement-on-us-uk-data-protection-discussions.

15. Manyika, James; Lund, Susan; Bughin, Jacques; Woetzel, Jonathan; Stamenov, Kalin; and Dhingra, Dhruv. "Digital globalization: The new era of global flows." McKinsey Global Institute Report, McKinsey & Company, February 2016, https://www.mckinsey.com/business-functions/digital-mckinsey/our-insights/digital-globalization-the-new-era-of-global-flows.

16. Castro, Daniel, and McQuinn, Alan. "Cross-Border Data Flows Enable Growth in All Industries." The Information Technology and Innovation Foundation, February 2015, http://www2.itif.org/2015-cross-border-data-flows.pdf?_ga=1.8208626.1580578791.1473954628.

17. Porter, Michael E., and Heppelmann, James E. "How Smart, Connected Products Are Transforming Competition." Harvard Business Review, November 2014, https://hbr.org/2014/11/how-smart-connected-products-are-transforming-competition.

18. Gandel, Stephen. "Lloyd's CEO: Cyberattacks cost companies $400 billion every year." Fortune, January 23, 2015, http://fortune.com/2015/01/23/cyber-attack-insurance-lloyds.

19. "Child Pornography: Model Legislation & Global Review." Child Pornography: Model Legislation & Global Review, International Centre for Missing & Exploited Children, https://www.icmec.org/child-pornography-model-legislation.

20. "A Cloud for Global Good." A Cloud for Global Good, Microsoft, 2016, https://news.microsoft.com/cloudforgood/policy/briefing-papers/responsible-cloud/technology-fraud-online-exploitation.html.

21. Trapasso, Ed. "Microsoft, Accenture and WSP Environment & Energy Study Shows Significant Energy and Carbon Emissions Reduction Potential from Cloud Computing." Accenture, November 5, 2010, https://newsroom.accenture.com/industries/systems-integration-technology/microsoft-accenture-and-wsp-environment-energy-study-shows-significant-energy-and-carbon-emissions-reduction-potential-from-cloud-computing.htm.

22. "Countries adopt plan to use Internet in implementation of Sustainable Development Goals." Department of Economic and Social Affairs, United Nations, December 16, 2015, https://www.un.org/development/desa/en/news/administration/internet-for-sdgs.html.

23. "The 2015-16 Affordability Report." Digital Empowerment Initiative for Eastern Africa, February 29, 2016, https://www.deiea.net/the-2015-16-affordability-report.

24. "Top 50 countries with the highest internet penetration rates," Internet World Stats, http://www.internetworldstats.com/top25.htm.

25. The 2015-16 Affordability Report." Digital Empowerment Initiative for Eastern Africa, February 29, 2016, https://www.deiea.net/the-2015-16-affordability-report.

26. "The skills gap in U.S. manufacturing: 2015 and beyond." Deloitte, 2015, http://www.themanufacturinginstitute.org/~/media/827DBC76533942679A15EF7067A704CD.ashx.

27. Microsoft Cloud Skills Report: Closing the Cloud Skills Chasm." Microsoft Business, Microsoft, https://msenterprise.global.ssl.fastly.net/wordpress/2017/03/UK-Cloud-Skills-Report.pdf.

28. Chen, Li-Kai; Grant, Andrew; and Mourshed, Mona. "The $250 billion question: Can China close the skills gap?" McKinsey Global Institute Report, McKinsey & Company, June 2013, https://www.mckinsey.com/industries/social-sector/our-insights/the-250-billion-question-can-china-close-the-skills-gap.

29. "Crunched by the Numbers: The Digital Skills Gap in the Workforce." Burning Glass, March 2015, http://burning-glass.com/wp-content/uploads/2015/06/Digital_Skills_Gap.pdf

30. "Overview." The World Bank, Global Findex, Washington, D.C., 2016, http://www.worldbank.org/wdr2016.

31. "New world report shows more than 1 billion people with disabilities face substantial barriers in their daily lives." World Health Organization, June 9, 2011, http://www.who.int/mediacentre/news/releases/2011/disabilities_20110609/en.

32. Woetzel, Jonathan; Madgavkar, Anul; Ellingrud, Kweilin; Labaye, Eric; Devillard, Sandrine; Kutcher, Eric; Manyika, James; Dobbs, Richard; and Krishnan, Mekala. "How advancing women's equality can add $12 trillion to global growth." McKinsey Global Institute Report, McKinsey & Company, September 2015, https://www.mckinsey.com/global-themes/employment-and-growth/how-advancing-womens-equality-can-add-12-trillion-to-global-growth.

33. McCarthy, Douglas, and Klein, Sarah. "The Triple Aim Journey: Improving Population Health and Patients' Experience

of Care, While Reducing Costs." The Commonwealth Fund, July 22, 2010, http://www.commonwealthfund.org/publications/case-studies/2010/jul/triple-aim-improving-population-health.

34. "Health expenditure, total (% of GDP)." World Health Organization Global Health Expenditure database, The World Bank, 2017, https://data.worldbank.org/indicator/SH.XPD.TOTL.ZS/?end=2014&start=1995&view=chart.

35. Lawry, Tom. "Saving lives with data." Enterprise, Microsoft, September 13, 2016, https://enterprise.microsoft.com/en-us/industries/health/saving-lives-with-data.

36. Lawry, Tom. "Saving lives and money with machine learning." Enterprise, Microsoft, September 28, 2016, https://enterprise.microsoft.com/en-us/industries/health/saving-lives-money-with-machine-learning.

37. "365mc Unveils World's First AI-assisted Liposuction System 'M.A.I.L System' Developed in Collaboration with Microsoft." Business Wire, Yahoo! Finance, October 18, 2017, https://finance.yahoo.com/news/365mc-unveils-worlds-first-ai-120000940.html.

38. "Volpara: Applying New Zealand Technology to Solving Global Clinical Problems." Customer Stories, Microsoft, August 15, 2017, https://customers.microsoft.com/en-us/story/volpara-health-azure-cloud-service-powerbi-en.

39. "The vision for AI in Health." Democratizing AI in Health, Microsoft, http://democratizing-ai-in-health.com.

40. “Soddo Christian Missionary Hospital Case Study.” Customer Stories, Microsoft, February 7, 2017, https://enterprise.microsoft.com/en-us/customer-story/industries/health/soddo-christian-missionary-hospital-case-study.

41. “Virtual Hospital improves patients’ healthcare access, dramatically cuts costs.” Customer Stories, Microsoft, September 29, 2017, https://customers.microsoft.com/en-US/story/helsinki-university-hospital-health-office-365.

42. Microsoft in Business Team. “How Children’s Hospital LA saves eyesight with telemedicine.” Enterprise, Microsoft, July 11, 2017, https://enterprise.microsoft.com/en-us/articles/blog/microsoft-in-business/how-childrens-hospital-la-saves-eyesight-with-telemedicine.

43. BizSpark Online. “DERPi leverages Azure’s hosting capabilities and flexible scaling to fuel their new diabetes intervention and prevention application, HiPTT.” TechNet, Microsoft, July 27, 2016, https://blogs.technet.microsoft.com/bizspark_featured_startups/2016/07/27/derpi-leverages-azures-hosting-capabilities-and-flexible-scaling-to-fuel-their-new-diabetes-intervention-and-prevention-application-hiptt-2.

44. “Ground-breaking service for millions with emotional and psychological problems is launched from the cloud.” Enterprise, Microsoft, April 201, 2015, https://enterprise.microsoft.com/en-us/customer-story/industries/health/shezlong.

45. “Vivli Collaborates with Microsoft and BlueMetal on a New Global Clinical Trial Data Platform to Accelerate Scientific Discoveries for Human Health.” Business Wire, July 12, 2017, http://www.businesswire.com/news/home/20170712005213/en/

Vivli-Collaborates-Microsoft-BlueMetal-New-Global-Clinical.

46. PR Newswire. “BC Platforms Launches Unique Global Genomics And Clinical Data Analytics Resource.” The Street, September 13, 2017, https://www.thestreet.com/story/14303225/1/bc-platforms-launches-unique-global-genomics-and-clinical-data-analytics-resource.html.

47. “EU Cloud in Health Advisory Council Offers Recommendations on Health Data.” HealthTech Wire, HiMSS Europe, October 5, 2017, http://www.himssinsights.eu/eu-cloud-health-advisory-council-offers-recommendations-health-data.

48. “New IDC Forecast Sees Worldwide Big Data Technology and Services Market Growing to $48.6 Billion in 2019, Driven by Wide Adoption Across Industries.” Business Wire, November 9, 2015, http://www.businesswire.com/news/home/20151109005070/en/New-IDC-Forecast-Sees-Worldwide-Big-Data.

49. “Report IV: Small and medium-sized enterprises and decent and productive employment creation.” International Labour Conference, 104th Session, International Labour Office — Geneva, 2015, http://www.ilo.org/wcmsp5/groups/public/---ed_norm/---relconf/documents/meetingdocument/wcms_358294.pdf.

50. Delgado, Rick. “Study Shows Small Businesses Are Doubling Profits by Switching to Cloud Computing.” Business.com, February 22, 2017, https://www.business.com/articles/study-shows-small-businesses-are-doubling-profits-by-switching-to-cloud-computing/